S0-BXA-003

ONLINE COMMUNITIES

HUMAN/COMPUTER INTERACTION

A Series of Monographs, Edited Volumes, and Texts

SERIES EDITOR
BEN SHNEIDERMAN

Directions in Human/Computer Interaction
Edited by Albert Badre and Ben Shneiderman

Online Communities:
A Case Study of the Office of the Future
Starr Roxanne Hiltz

Human Factors In Computer Systems
Edited by John Thomas and Michael Schneider

Human Factors and Interactive Computer Systems
Edited by Yannis Vassiliou

ONLINE COMMUNITIES

A Case Study of the Office of the Future

Starr Roxanne Hiltz

Upsala College

Ablex Publishing Corporation
Norwood, New Jersey 07648

Printed in the United States of America.

Library of Congress Cataloging in Publication Data

Hiltz, Starr Roxanne.
Online communities.

(Human/computer interaction)
Includes bibliographical references and index.
1. Interactive computer systems. 2. Communication in science. 3. Computer networks. I. Title.
II. Series.
QA76.9.I58H54 1983 384 83-6076
ISBN 0-89391-145-3

Ablex Publishing Corporation
355 Chestnut Street
Norwood, New Jersey 07648

CONTENTS

LIST OF TABLES

PREFACE

New technology offers a challenge and opportunity to system developers and entrepreneurs. As computers and communications matured, it became possible to create electronic mail, teleconferencing, and office automation. Ambitious designers and adventurous entrepreneurs created novel products, which are still being refined, to support human interaction by computer. Pioneering efforts are demanding because the potential for failure is high, there are no human role models to follow, and the lack of support from colleagues and competitors can be demoralizing.

New technology offers a similar challenge and opportunity to academic and industrial scientists. Those involved in advanced research often face an existential loneliness which demands a strong character to endure. When academic research crosses traditional disciplinary boundaries or introduces novel paradigms, colleagues become more distant and competitors may feel obliged to use the opportunity for criticism.

Starr Roxanne Hiltz has been a pioneer in exploring the application of sociological techniques to the study of group interaction through computer conferencing. She has created novel paradigms of research which are unsettling to traditional computer scientists and she has used technology which is unfamiliar to traditional sociologists. Her dedication for so many years and her courage in pursuing this direction have yielded numerous benefits.

This book presents the details of her two year study of how groups of knowledge workers communicated, cooperated, and cajoled. Dr. Hiltz shows how these groups succeeded and how they failed. She offers practical guidance to system designers, organizational managers, and individuals who will use computer conferencing.

The statistical data and the anecdotal evidence supported some of my beliefs and refined my understanding of human-computer interaction. Evidence of increased interaction among computer conferencing participants was encouraging. Clarification of the effectiveness of computer conferencing for planning and collaboration was promising. The negative results concerning conflict resolution are an important warning for potential users.

Hiltz's work makes clear that the design of the technology influences and is influenced by the environment of its use. By selecting participants and structuring the topic, electronic mail becomes a new social form—computer conferencing. The permanent record documents who said what and when, and allows review by new participants. This historical trace compels participants to more carefully consider their comments.

These results will shape the next generation of office automation services and influence training, installation, and operating procedures. Equally important are the impacts that this book will have on the research methodology for sociologists and computer scientists. Sociologists should see that they have a golden opportunity, not only to study group interaction, but to influence the evolution of this vital technology. Computer scientists and system developers should see the power of collaboration with social scientists in refining the design and implementation of their technological innovations.

Ben Shneiderman
Series editor

FOREWORD

What is an "office"? Usually one thinks of it as a place, with desks and telephones and typewriters. In thinking about the office of the future, one must instead think of it as a communications space, created by the merger of computers and telecommunications. The "office workers" and the computers which support them will be located in many different places. Some will work from computer terminals in their homes. Most of the work they do—communicating with one another, creating and analyzing information, making decisions and implementing them—will take place "online," mediated and supported by a computer network.

How do people react to spending much of their life online? What determines whether they will enthusiastically incorporate a computer-mediated communication system into their daily lives, or reject it? What do they like and dislike about such systems? What impacts does it have on the organizations for which they work, and on their productivity? The prototype systems of today can help us answer these important questions for tomorrow.

The first computer-mediated communication systems were developed and used in the 1970's. Today there are hundreds of systems. Most are simple "electronic mail" systems. Others provide a variety of capabilities to support individual managers who are composing a document or working with a data base; dyads or small groups who are exchanging messages; and large groups which are working together on an extended project. Several of the systems which fall into the latter category are called "computerized conferencing" systems. They can be seen as prototypes of the communication-information systems that will link the "offices" of the future. Besides EIES (the Electronic Information Exchange

System, which serves as the example for this case study), other current examples include HUB, designed at The Institute for the Future as an outgrowth of their PLANET conferencing system; OICS (Office-Information-Communication System) designed by Bell Northern Research in Canada; Augment (formerly called NLS and offered on the TYMSHARE network); CONFER (University of Michigan) and COM, a Swedish conferencing system.

This is a study of several scientific research communities which used EIES for a period of about two years to enhance their communications and carry out cooperative tasks. Though it focusses on one particular system, it was designed to yield some data that make possible direct comparisons with the results of studies of other computer-mediated communication systems. Included are an examination of the determinants of acceptance of this new form of communication; user reactions and preferences related to specific system features and design choices and how these change with experience; and changes in communication patterns, work patterns, and productivity-related measures as a result of using the system.

The case study should appeal to those interested in the applications and social impacts of computer-mediated communications systems and their design and evaluation. It is also relevant for those interested in the role of communication in scientific research specialties in general, and the relationship between technological innovation and social change in general. Specifically, it is a case study of a social invention, aimed at:

- Managers who must decide whether to implement such systems in their offices
- Students of communications, office automation, the design of interactive computer systems, and the social impacts of technology.

William Whyte (1980:5) defines a social invention as:

> . . . a new and apparently promising strategy designed to solve some persistent and serious human problems. It may take the form of a new organizational structure or a new set of interorganizational relations. It may involve a new set of procedures for shaping human interactions and activities and the relations of humans to the natural and human environment.

Computer-mediated communication systems, if they live up to the hopes of their designers, are a social invention in all of these senses. This study is an attempt to describe the nature and impacts of one such case history in social invention, the Electronic Information Exchange System (EIES), which had as its objective the enhancement of communication and productivity within scientific research communities.

Starr Roxanne Hiltz

ACKNOWLEDGEMENTS

This book grew out of a research project financed by the National Science Foundation. Much of it is adapted from the final report for the project (MCS–77–27813). The opinions and conclusions are solely those of the author and do not necessarily represent those of the National Science Foundation.

The author is indebted to Murray Turoff for coauthoring the sections describing the EIES system and for his suppport and encouragement for this study at all stages. Mary Anne Solimine served as a research assistant, supervising the distribution, coding, and tabulations of questionnaire responses. Without her diligent efforts, the study would not have been possible. Anita Graziano provided invaluable administrative support at NJIT. Dutchess Brooks, Lisa Fetterman, Joanne Garafalo, Sonia Khalil, Christine Naegle, Diane Price, Ann Marie Rabke, Marion Whitescarver, and Margaret Wnorowski provided assistance with coding, and data entry and editing tasks. Larry Landweber was most cooperative in providing access to the Theory Net Group. Alan Leurck, Thomas Moulton, and Sanjit Chinai are among those at NJIT who prepared statistical data from information on users recorded by the system monitor.

Among those who have made helpful contributions to the project are Diana Crane, Kenneth Johnson, Peter and Trudy Johnson-Lenz, Elaine Kerr, Ian Mitroff, Nicholas Mullins, Ronald Rice, Julian Scher, and Barry Wellman. Initial interest in the sociology of science was inspired by the work of Robert Merton, who of course bears no responsibility for the directions taken by his student. Thomas Kuhn and the members of the sociology of science seminar at the Institute for Advanced Study in 1976–77 helped in the process of turning this vague interest into a concrete research proposal. Fred Weingarten, formerly of the

National Science Foundation, and W. Richards Adrion, currently with the National Science Foundation, have the author's gratitude for their support of the research. Last, but certainly not least, Ben Shneiderman provided many helpful editorial suggestions for the revision of the final manuscript.

PRIOR PUBLICATION

Portions of the following chapters have appeared as follows:

Chapters 1 and 4: "The Evolution of User Behavior in a Computerized Confererencing System," (coauthored by Murray Turoff) Communications of the ACM, Nov. 1981.

Chapter 3: "The System Is as the User Group Does," Proceedings of the Amer. Soc. for Info. Sci., Fall 1981.

Chapter 4: "Office Augmentation Systems: The Case for Evolutionary Design" (with Murray Turoff), Proceedings of the Fifteenth Hawaii International Conference on System Sciences, Jan. 1982.

Chapter 5: "Human Diversity and the Choice of Interface: A Design Challenge" (with Murray Turoff), Proceedings of the ACM–SIGSOC Joint Conference on Easier and More Productive Use of Computer Systems, University of Michigan, May 1981.

Chapter 7: "Impact of a Computerized Conferencing System Upon Use of Other Communication Modes," Proceedings of the Int. Conf. on Computer Communications, London, Sept. 1982.

Chapter 7: "The Impact of a Computerized Conferencing System on the Productivity of Scientific Research Communities," Behavior and Information Technology, Vol. 1, 1982:185–195.

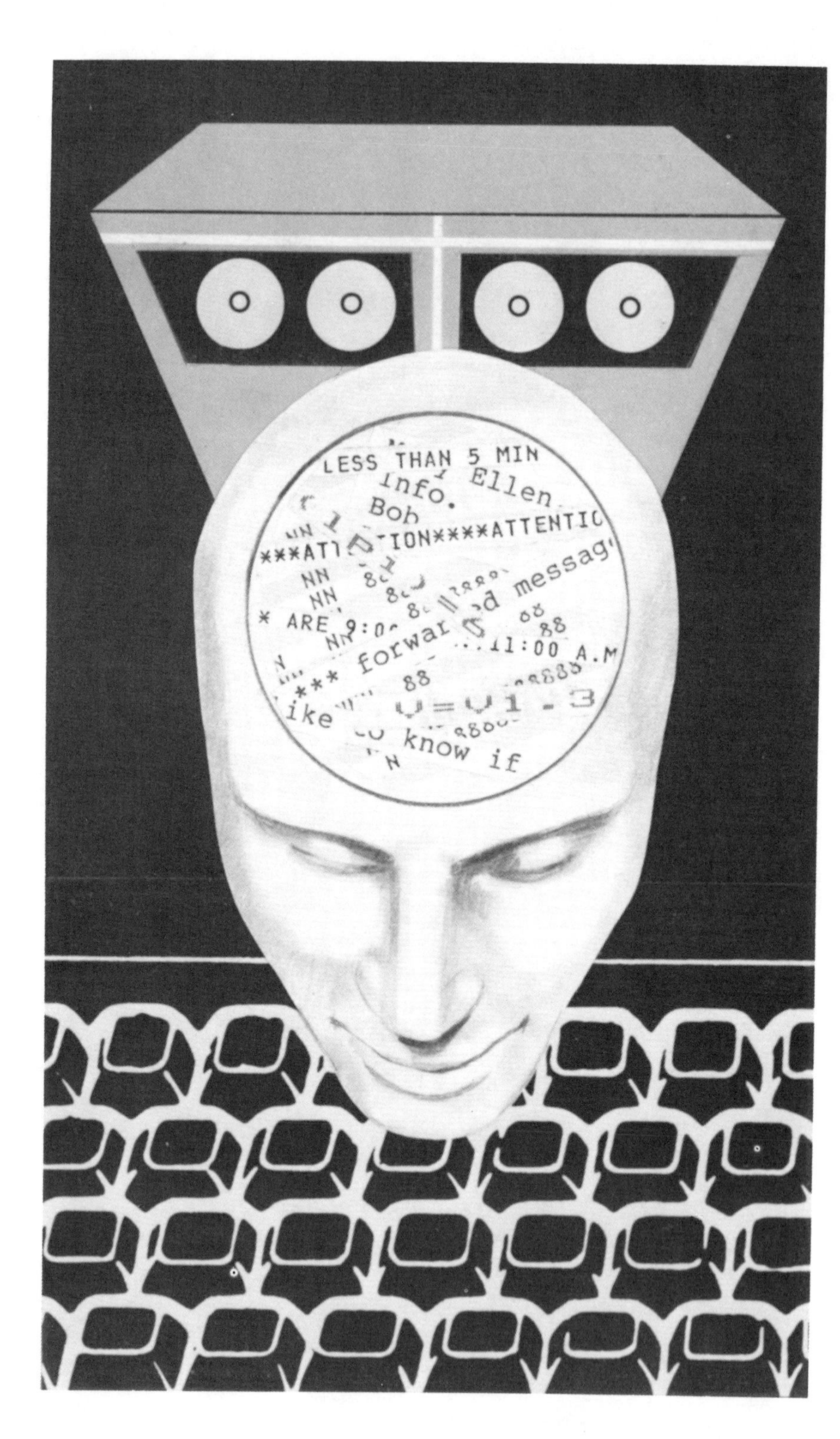
LESS THAN 5 MIN
Ellen
Info.
Bob
11:00 A.M
know if

1.

THE CASE STUDY

INTRODUCTION

In the industrial revolution, machines were used to augment human physical labor. Analysts ranging the ideological spectrum from Adam Smith to Karl Marx recognized that the introduction of machines resulted in a fundamental change in economic and social organization. Workers were not always happy about these changes; the Luddites smashed machines. Now we are undergoing another fundamental economic and social change. Computers and telecommunications are being used to augment human "knowledge work." First computers were used to manage information in the office. Now computer-mediated communication systems are being designed to support professionals, managers, and clerical workers.

Computer-mediated communication systems use a computer to create, store, process, and distribute communications among groups of humans. The users can exchange ideas and information on a regular basis without having to be in the same place or to communicate at the same time. (For overviews of the technology and its impacts, see Hiltz and Turoff, 1978; Johansen, Vallee, and Spangler, 1979; and Kerr and Hiltz, 1982). Electronic mail systems treat each item as a separate "letter" or file. Some systems also provide structures for permanent transcripts of group discussions on various topics; these are usually called "computerized conferencing" systems ("CC"). Among the fundamental characteristics of CC as a mode of communication are that:

- Communication takes place through a computer terminal by typing and reading. Both cognitive and social-emotional exchanges tend to be different than face-to-face communication (See Hiltz, Johnson, Aronovitch, and Turoff, 1980).

- Communication is "asynchronous"; sending and receiving may occur seconds apart, or days or even years apart.
- The computer stores communications and information, allowing retrieval by attributes such as topic. The user can also filter communications, deciding whether, when, and how thoroughly to choose to read items from the mass of accessible material.
- The computer can be programmed to provide a variety of communication structures and services, such as tabulation and display of votes, or analysis and display of information according to a format specified by a particular individual or group (See Hiltz and Turoff, 1978, Chapter 9).

This is a case study of the users of a computerized conferencing system called EIES (Electronic Information Exchange System, based at the New Jersey Institute of Technology and designed by Murray Turoff). It examines its acceptance by and impacts upon a particular professional and technical population: groups of scientists working within the same research specialty. In order to improve the generalizability of the findings to other systems and types of users, the data collected were coordinated with those being gathered for other projects occuring at about the same time. The dual purposes of this case study are to convey an "in depth" understanding of reactions to and effects of this CC system on the scientific communities which used it, and to assess the implications of these findings for future implementations of similar systems. It might be noted at the outset that there are no documented cases of users smashing their terminals; but several reported that they occasionally felt like it.

EIES was originally designed to enhance communication within geographically dispersed "small research communities" of scientists, "conceived as groups of 10 to 50 individuals sharing an interest in a scientific or technological problem area" (NSF 76–45:3). EIES provides a message system which enables members to send private communications to individuals or groups on the system, "conferences" which build up a permanent transcript on a topic of discussion, and notebooks where text processing features may be used to work on jointly authored reports. It also provides the capability to handle unique kinds of information or knowledge bases, or to change the interface, or conduct a controlled experiment. For example, one of the scientific groups in this study had capabilities designed for them to produce an "electronic journal." Another group had software designed to facilitate inquiry-response exchanges that followed a selective tree-like structure rather than the linear transcript structure of the regular conference system (See Johnson-Lenz, 1981a).

THE CASE STUDY IN CONTEXT

What does it mean when we say that this is "a case study of the office of the future"? How does this case study fit into the social and economic context of office work in general? The elements that we need to consider are the nature of "knowledge" or "office" work, the place of scientific, engineering, and other professional and technical workers in the office labor force, the role of communications in offices, and the extent to which the particular system chosen can yield generalizable results.

"Knowledge" or "office" workers include all those whose primary function is to work with information: creating it, transforming and communicating it, using it, filing and retrieving it. Post-industrial societies are becoming "information" or "knowledge based" economies in the sense that a majority of their workforce now consists of "knowledge workers." As of 1979, the largest category was "clerical" workers (18% of the workforce), followed by professional and technical workers (over 15 million, comprising 16% of the workforce), and managers and administrators (11%).[1]

Clerical workers will eventually follow their bosses (the professional and technical, managerial and sales workers) online; but as "support" personnel, they are not the "leading edge." Our case study thus focusses upon the largest category of knowledge workers who are likely to be among the early users of computer-mediated communication systems, the professional and technical workers. Within this category, it chooses the scientists and engineers, who are considered the "key" to the future transformation of post-industrial societies:

> The central occupational category in the society today is the professional and technical . . . The most crucial group in the knowledge society, of course, is scientists, and here the growth rate has been the most marked of all the professional groups . . . [They] form the key group in the post-industrial society. (Bell, 1973:137, 216, 17).

The second premise in the assumption that this case study is a look at "the office of the future" is that computer-mediated communication systems will be the central technological innovation characterizing the new type of office work. This assumption is based upon the recognition that communication is the most frequent activity of office workers, and upon a view of organizations as primarily communication systems. As

[1]These figures were obtained by dividing the total reported workforce (96,945,000) by the numbers shown for the various categories. Many but not all of the "sales" workers (6,163,000, or 6%) are also primarily engaged in information exchange and communication, rather than the physical handling of "things." (Bureau of the Census, 1980: 418–420).

summarized by others who have previously looked at the office of the future,

> Studies have shown that the most frequent thing that office workers do is to communicate information . . . estimates of the amount of time a knowledge worker spends in the communication activity range from 50% to 90%. . . . The changes in the office of the future can be understood and consequently managed if the office and the organization (a collection of offices) are conceptualized as a complex communication system. (Uhlig, Farber, and Bair, 1979: 13, 23, 231)

The third element in the choices made for this case study is the particular system selected. To what extent would results be similar for other implementations of the technology? The assumptions that have been made are that:

- The system should not be so badly designed that prospective users are alienated by its particular interface, lack of features, or lack of reliability.

Feedback from users (see Chapters 2 and 5) indicates that this was not the case.

- The system should represent a range of the new capabilities, rather than just a narrow subset.

As described briefly above and in more detail below, EIES at the time of the study did include a range of types of features that might be used by office workers: text editing, document distribution, message exchange, conferences, etc. At one extreme, it might be assumed that systems which include only a subset of the capabilities of EIES might produce only a subset of the impacts observed for this system. At the other extreme, the newest and most powerful of the systems that will become available in this decade might be assumed to produce "more" of the results we have observed. For example, at the time of this study, most EIES users had only "dumb" terminals operating at 30 characters per second. As we begin the 80's, more and more office workers (and EIES users) are getting their own "personal computers" or microcomputers which can serve as intelligent terminals and greatly increase the speed with which information is displayed and communicated.

In short, our justification for the particular system chosen for the case study is that it is a reasonable representation of the current "state of the art."

OVERVIEW OF THE CHAPTER

This introductory chapter reviews previous findings about scientific communication which formed the initial basis for many of the variables examined in this study. It includes an overview of the EIES system, characteristics of the scientific user groups, and a description of the evaluation methods. It also discusses the problem of the extent to which the findings for this case study may be generalizable to other systems and other types of knowledge workers. In this context, a number of other case studies are described, which will be used throughout the book to determine if the findings of this study are comparable to those for other prototypes of the office of the future.

BACKGROUND: THE ORIGIN OF THE STUDY

The Division of Science Information (now the Division of Information Science and Technology) of the National Science Foundation issued a program announcement in 1976 inviting proposals for "operational trials" of EIES. Four groups were chosen to participate beginning in late 1977; three final groups were chosen in 1978. In addition, several other groups made use of the EIES system with DIST permission, but without DIST support.

The official objectives of the Operational Trials program were:

- To test the hypothesis that EIES can enhance the effectiveness of individuals belonging to such a community.
- To accumulate practical experience with EIES by the members of such a community.
- To gain deeper insight into the relationship between communication processes and the progress of science and technology. (NSF–76–45:3)

The Division of Mathematical and Computer Research funded a study by this author to conduct a cross-group assessment of the impact of the use of EIES, with the following objectives:

- Feedback to the designers on user reactions to system features;
- Isolation of the factors accounting for low vs. high use;
- Identification of impacts of the system on those individuals and groups who are heavy users.

We hoped to make the study comparative across systems. One other scientific user community on MACC-Telemail, theoretical computer scientists, did agree to participate. However, response rates from that group were fairly low. Indirect comparisons to the PLANET and NLS systems are made possible by using some of the same questions that had been employed in earlier studies of these systems.

This book is a reworking of the final technical report on the study (Hiltz, 1981). Tables documenting non-relationships and observed relationships of limited interest have been eliminated, along with many of the methodological details. Comparison with the findings of similar studies and discussions of the implications of the findings have been added.

THEORETICAL FRAMEWORK AND INITIAL HYPOTHESES

The analytical model begins with several "input" or independent variables: characteristics of the individual users, the scientific user groups, the tasks they undertook on the system, and the system itself. This framework was initially developed by Jacques Vallee and his colleagues at the Institute for the Future (Vallee et al., 1974:22) in their evaluative work on the PLANET computerized conferencing system. Characteristics of individuals which were measured include skills (such as typing and previous computer experience), initial attitudes toward the system they were invited to use, pre-existing patterns of communication and exchange with other scientists in the specialty, and access to computer terminals. Among the important characteristics of the group are its size, cohesiveness, leadership, and the task it is trying to accomplish through the computerized conferencing system. Important system characteristics include ease of learning, quality of documentation, "friendliness" of its interface, and the capabilities which it offers. (See Kerr and Hiltz, 1982, Chapter 2, for a complete review of system characteristics and their relative importance for user acceptance).

As a result of the interplay of individual, group, and system attributes, individuals choose whether or not to use the system. Some become dropouts; some become "addicts" who spend several hours a day working and communicating online. Through systematic feedback, the system itself undergoes change. The individuals and user groups also change, as a function of how much they use the system. This study collected data over time which could be used to track this complex process. Subsequent chapters discuss and review the individual, group, and system attributes which influenced the impact of EIES.

The sections which follow summarize the previous research which aided in conceptualizing the potential effects of EIES use upon scientific research communities. They represent the background for and justification of scientific research communities as the initial population for a study of the potential impacts of computerized conferencing systems.

CHARACTERISTICS OF SCIENTIFIC RESEARCH COMMUNITIES

Scientific Communication. Scientific specialties consist of a set of scientists who engage in research along similar lines and who communicate often (Hagstrom, 1970: 91–92). As Chubin (1975: 1) has pointed out, "disciplines form the teaching domain of science, while smaller intellectual units (nestled within and between disciplines) comprise the research domain." Such specialties have sometimes been called "invisible colleges" of scientists (Price,1963; Crane, 1972; Griffith and Mullins, 1972) and have been seen as the social location of technical, cognitive, and ethical norms (Mulkay, 1972; Mitroff, 1974a) and as internally stratified on the basis of productivity (Cole and Cole, 1973).

Geographically dispersed networks of scientists working in the same specialty area can be viewed as the prototypical "production organization" of science, in which the "product" is scientific knowledge, and the social organization depends largely upon the communication system. Formal and informal communication channels serve not only to direct and redirect efforts to "important" areas and fruitful methodological tools, but also to reinforce shared norms and theories and to allocate rewards in the form of recognition.

Cole and Cole (1973:16) describe the importance of communication in science as follows:

> Scientific advance is dependent on the efficient communication of ideas. The communications system then is the nervous system of science; the system that receives and transmits stimuli to its various parts.

Scientific communication processes have not changed in decades, except that in many disciplines, an exponential growth has slowed down the process and lengthened the time between the completion of a research project and its publication in a journal. Summarizing the results of a series of studies of scientific communication in psychology, which is similar to many other disciplines, Garvey and Griffith (1971: 354, 355) conclude that the scientist relies heavily on informal discussion, small meetings, and exchange of drafts and preprints. This interchange helps researchers to keep abreast of current activities and community views on the value of specific research problems. The journal article, by the time it is published, lags so far behind the research frontier that its functions are mainly to inform scientists in other specialties, and to allocate recognition for scientific achievement.

Increasingly, there have been calls for improving scientific communication and information dissemination. Many of these have focused on the information storage, processing, and networking capabilities of the computer to provide assistance.

Some of the suggested innovations deal with the formal communication channels, the professional meeting and the journal. There were an estimated 100,000 journals published in 1980; something must be done to decrease the costs and increase the efficiency of dissemination of "published" results. Selective dissemination of articles only to consumers who peruse computerized abstracts and order a copy of the full paper has been one answer; another has been more efficient, computer-assisted publishing procedures (See Rhodes and Bamford, 1976).

Another approach has been to make scientific information, particularly in the form of data bases and bibliographic files, directly available to researchers through an online, interactive computer system. One example of this is the NIH-EPA Chemical literature on a central computer which can be accessed from telephone-coupled computer terminals anywhere in the world. The user searches and retrieves information and performs data analyses on these files through interactive programs. (See Heller, et al., 1977). Scores of abstracting services have also been computerized to allow users to search through these files using combinations of key words.

However, the informal, pre-publication communication within scientific specialties is also crucial to increasing scientific productivity. Recognizing this, the Division of Science Information of the National Science Foundation financed the building and field testing of EIES as a computer-mediated communication system designed specifically to meet the needs of networks of geographically dispersed scientists.

Would this new capability actually be used to share and cooperate? Or would the new, potentially more efficient, means of exchanging information and ideas seem too "inhuman" or too difficult to use? These issues are treated in Chapter 2.

Impacts on the Resolution of Scientific Controversies. We were especially interested in the impact of CC upon research specialties experiencing sharp theoretical controversies, with the competing theories each having their adherents. Studies by distinguished analysts of science such as Kuhn, Merton, and Feyerabend have established that controversies are a perpetually recurring, if not permanent, feature of science. Such studies also establish that controversies are a vital feature of science since they introject fresh points of view and challenge established beliefs. Some hypotheses will clash sharply, since they are frequently based on different ideologies (see Robbins and Johnson, 1976). Controversies are likely to be endemic in new, interdisciplinary specialties which have never developed a shared paradigm; the EIES groups tended to be in this pre-paradigmatic stage of development. (See Kuhn, 1970, for an examination of the nature of scientific paradigms and stages of development; see Mullins, 1972, for a case study of the emergence of a new specialty.)

One can imagine the emergence of a new scientific research paradigm as a kind of Hegelian dialectical process. A new theory or method arises to challenge the existing dominant approaches. There may be a period of increased controversy as the two sides argue and clarify their differences. Then the controversy is resolved by synthesis of opposing viewpoints. A full cycle might take more than the eighteen to twenty-four months groups spent on EIES. Thus, this study looks for both parts of the hypothesized process: the clarification of controversies and their resolution (see Chapter 6).

Would the availability of this new channel of communication foster the fruitful exchange of alternative viewpoints? Or would EIES become a kind of electronic soapbox, with the pontifications of individuals largely ignored by the other members of the community?

Impacts on Communication and Productivity. Existing scientific communications structures are either very slow (printed journals), very fitfull and expensive (yearly conferences or special meetings), or very exclusive (personal letter, personal visit, or telephone call). Computerized conferencing enables users to keep in constant communication with one another and to exchange ideas and findings on a daily-to-weekly basis, sending and receiving such materials at their own convenience. It could increase the amount and timeliness of the raw materials (information and ideas) used in the scientific process and thus increase the productivity of scientists. Alternatively, do users become so enamored of their new technological toy that they lose sight of it as a means to an end and become less productive? Do they get so overloaded with new information and ideas that they cannot assimilate them? (see Chapter 7).

The Sociometric Structure of Specialty Areas. Another area of inquiry is the impact of conferencing systems upon the size of and status differentiations within the research communities. This set of issues is treated in several chapters. Will computerized conferencing condense the research specialty into a smaller core group, with those not in the system more completely cut off? Or, will the increased ease of communication within this core facilitate expansion through circulation of printed output and invitations to "observers" or "visitors" to occasionally take part? Would time be freed to do more letter writing and manuscript circulation to more people, and/or the facilitation of specialized face-to-face conferences to which a general invitation is extended? Or will use of the network take so much time and energy that off-line communications and relationships suffer?

Scientific research communities are not only communication networks, but are also stratified social systems which allocate prestige and opportunities. For example, as Price and Beaver (1966:101–117) describe their concept of invisible colleges:

> In each of the more actively pursued and highly competitive of the sciences there is an "in-group." The people in such a group claim to be reasonably in touch with everyone else who is contributing materially to research in this subject . . . They commute between one center and another, and they circulate preprints to each other and they collaborate in research . . . They may . . . control personal prestige and the fate of new scientific ideas, and intentionally or unintentionally they may decide the general strategy of attack in an area.

Two interesting inadequacies of the "invisible college" structure are immediately obvious. First, for those who are "in," the existing communications network is so time-consuming, sporadic and slow, that only a few of the many questions, answers and comments that might fruitfully be exchanged actually are. Second, what about those potentially productive scientists who are "out"? An analysis of productivity patterns of chemists (Reskin, 1977:441) suggest that "collegiate recognition is particularly important for chemists in contexts that that do not stress scholarly publication."

A computerized conferencing system might make the exchange within "in" groups more effective. It also could allow the rapid formation of communities that do not now exist. A group of younger unknown researchers could form their own peer group independent of the "established" in-group. Moreover it could allow new modes of interaction between "elites" and "newcomers" (see Mulkay, 1976, for one view of current relationships).

Thus an issue of interest is which scientists can be most aided by such a system, those who are already part of a highly productive elite within a specialty, those who are currently cut off from extensive communications and cooperation with others in the field, or those at the "middle" levels? At present, the academic community is stratified, with those scientists located at the top universities having a much greater opportunity to be productive and gain recognition. They have more time, money and equipment for research, and their academic affiliation automatically includes them in an existing communication network. This is an example of what Merton (1968) calls the "Matthew Effect" in science, quoting from the Book of Matthew: "For unto everyone that hath shall be given, and he shall have abundance: but from him that hath not shall be taken away that which he hath."

Allison and Stewart (1974) have used cross sectional survey data to provide evidence that at least for chemists, physicists, and mathematicians, getting off on "the wrong foot" can severely lessen the opportunity to ever have the contacts and resources to be "productive" in terms of research. They summarize their findings as follows:

> The highly skewed distributions of productivity among scientists can be partly explained by a process of accumulative advantage. Because of feedback through recognition and resources, highly productive scientists maintain or increase their productivity, while scientists who produce very little produce even less later on. A major implication of accumulative advantage is that the distribution of productivity becomes increasingly unequal as a cohort of scientists ages. (p. 596)

A computerized conferencing system might increase equality of opportunity among researchers; those at small institutions might benefit most from the increased stimulation due to improved peer group communications. However, competitive pressures might make users unwilling to help out their peers. Would feelings of competitiveness increase, with the participants becoming unwilling to provide leads or information that would help their colleagues, perhaps at the expense of their own relative standing?

THE STRUCTURE OF EIES

EIES provides four general purpose structures for all its users:

- *Messages:* The delivery of messages to individuals and/or defined groups. This facility includes confirmations of delivery, a central message file, editing, retrieval, searching and resending, as well as historical analysis of message traffic by individuals.
- *Conferences:* Linear time sequential transcripts of group discussions on a particular topic with status information on readership. This facility includes voting, text searches, automatic delivery of new material to individual conferees and other communication support functions. Descriptions of open conferences are listed in a public conference, and an individual may join any number of conferences.
- *Notebooks:* A text composition and word processing space that may be private to an individual or shared among a group of users. Provides features for organizing and distributing documents as well as automatic notification of modifications.
- *Directory:* A membership directory containing both individuals and defined groups with self entered interest descriptions and numerous search options. A defined group may be treated as a single individual for purposes such as sending a message.

Messages are either private or group messages, and conferences and notebooks may either be private, group, or public. Private conferences and notebooks are controlled by an individual user who determines the participants. Group conferences and notebooks are controlled by defined groups on EIES, while public conferences or notebooks are available to anyone on the system for reading. Public notebooks have a defined set of authors (restricted writing), but anyone can read in them.

All the text items in the above subsystems are compatible and readily transferable, i.e., a message may be transferred into a conference comment or notebook page. All of the subsystems exist within the context of a single user interface that provides four different modes of user interaction. These are:

- *Menu Selection:* the user selects an option from the one page guide to the major EIES menus.
- *Command Driven:* all the menu selections are available as commands. In addition, approximately 200 advanced features not available in the menus can be utilized.
- *Answer Ahead and Command Streams:* The user can anticipate questions and answer ahead or trigger a sequence of operations. The EIES interface is fully predictable to the user and all commands are usable at any point in the interaction.
- *Self-defined Commands:* the individual user or a group coordinator can define commands unique to the individual or group. There are facilities for defining commands that will request necessary specifics at the time they are executed.

EIES also has a general purpose language (INTERACT) that can interpret any input stream from a user or from EIES as an executable program. INTERACT programs are stored in EIES text items. This capability allows tailoring of the interface and communication features by individuals or groups. Access to EIES programs is given by readership privileges on the text item in which it is stored.

EIES operates on a dedicated mini-computer—a Perkin-Elmer Interdata 7/32 with half a megabyte of core and two 300 megabyte disks. It supported up to 32 simultaneous users at the time of the study. EIES is implemented in Fortran, with modifications to the compiler and to the executive system. It is accessed either by a direct telephone call, or through the TELENET packet-switched network. TELENET had nodes in approximately 185 U.S. cities during the period of this study; the cost was $3.75/hour to connect to EIES from any of these nodes.

Within the basic structure of EIES are many specific system features. Many of these have been subjected to user evaluation. Table 1.1 provides a brief description of system features and indicates which evolved from late 1976 to 1980. The fact that the system was constantly evolving, partially as a result of feedback from this study, greatly complicated the problem of getting comparable data from users who joined the system at different times.

How would users cope with this rich, complex set of capabilities? Would any patterns emerge in terms of identifiable stages of learning and incorporation into work habits? What would users like and dislike about the features, the interface, and the documentation? These questions are examined in Chapters 4 and 5.

PARTICIPANTS IN THE STUDY: GROUP AND INDIVIDUAL CHARACTERISTICS

Four scientific communities (which became groups 30, 35, 40, and 45) began using EIES between October, 1977 and February, 1978. All four agreed to take part in this study. Face-to-face meetings were held with the Principal Investigators for these groups even before they formally submitted their proposals to NSF. The four initial user groups were all research communities which had as their primary goal the improvement of communication within the specialty area, rather than any particular task. Several other groups later joined the system and participated in either the full study or in use of some of the same survey questions. The Principal Investigator for each of the research communities was also responsible for conducting an evaluation of the project for that specific community. Listed below are the five research communities which participated fully in the study, a brief description, their average number of members, and references to available evaluations. Relevant results of these independent evaluations have been incorporated into this publication.

FUTURES—Group 30, "Futures Research Methodology," is composed of persons who have conducted planning, forecasting and similar studies, and are attempting to discuss and improve methodology in this area. As pointed out in the proposal submitted for this operational trial, "Since futures research methodologists come from a wide variety of backgrounds and disciplines, the channels of communication which would ordinarily be provided by a single professional society do not exist" (Martino, 1977:2). It was hypothesized that use of EIES would significantly enhance the rate of innovation and dissemination of fruitful new ideas in the field. These conditions and hopes are similar to those stated in the proposals for groups 35, 40, and 45. The average size of the Futures group was

TABLE 1 · 1 ***Brief Explanation of EIES Features***

Features in the Original Design

Private Messages: Can be sent to any individual or list of individuals. Confirmation of date and time of delivery is given.

Group Messages: Delivers the message to all members of a predefined group. No confirmations are provided, but sender can request status list showing who has received it.

Membership Directory: Self-entered short description and address for all groups and members. Specialized searches are incorporated.

Private Conferences: Any member may initiate and moderate a conference on any topic. Moderator has the right to choose participants and decide whether or not to advertise.

Group Conferences: Each Group has a permanent general conference to which all group members belong.

Public Conferences*: Conferences in which anyone on the system may read or write without having to be granted access.

Private Notebooks*: Each member has a notebook for composing and storing items. The owner of a notebook may give other members privileges to either read only or write as well. Owners may also establish read-only windows to portions of the notebook. New items as well as modifications of existing items are reported to all members in a notebook.

Group Notebooks*: Same features as private notebooks, but associated with all members of a group.

Public Notebooks*: Anyone on the system may read in a public notebook, but only the designated authors may write in the notebook.

Menus: The standard form of person-machine interface taught to new users via the written documentation they initially receive.

TABLE 1 · 1 *(cont.)*

Features in the Original Design *(cont.)*

Commands*: System-wide commands allowing the complete replacement of the use of menus and adding other unique capabilities outside those available through the menu.

Explanations: An online searchable file containing specific explanations of all system features.

Retrieval: The ability to recall any text item previously read and not yet deleted by a unique identifier. For messages this is limited to the last 30,000 sent on the system (about three months' traffic). For conferences or notebooks owners of these spaces delete items when they are outdated.

Searches*: Messages, conference comments and notebook pages may be searched by author, editor, dates, item identifier, free key words, full text, associations among items in either a nested or combination process.

Anonymity & Pen Names*: Any text item may be signed anonymously or with a unique secret pen name. Messages may be sent to pen names.

Synchronous Conferences: The ability to hold a conference when all members are online at the same time by supplying status indications of everyone's position in the conference at any time.

Voting*: The ability to choose any one or two of nine alternative voting scales that can be attached to a conference comment. The computer collects and displays the vote distribution for the members of the conference.

Direct Text Edits*: A line-oriented editor for use in the scratchpad, where individuals compose text items for entry into the system. Edits are accomplished immediately.

Copy, Get and See: Methods of indirectly referencing other items of text within a given text item or of transferring text items among messages, conferences and notebooks. In the case of 'See', the printout of an item is conditional on whether the receiver has already seen it.

TABLE 1 · 1 ***Brief Explanation of EIES Features (cont.)***

Evolved Features (Those added to the EIES system based upon feedback from users)
User Consultants: Volunteers who help others to learn to use the system and who also serve as information brokers on activities taking place on EIES. A number of special-purpose software features exist to facilitate the tasks of the user consultants.
CHIMO (newsletter): A summary of events taking place on EIES.
? or ??: Entering a ? or ?? as an answer to any question or choice on EIES results in a short or long explanation, respectively.
?word: Will retrieve an explanation of the "word" or system feature named from the explanation file.
SEN, ??? or LINK: Sending one-line messages which are delivered the next time the recipient does a carriage return, with or without confirmations or continuous exchange of one-liners with a group.
Defined Commands: Any user may define a sequence of operations or commands as an individually tailored command. Facilities exist for the more sophisticated user to make these conditional.
Indirect Edits: Edit commands stored within the text providing such things as centering, paging, text justification and tabulation. Indirect edits are executed at output time and are based upon the specifications the receiver has indicated for his or her terminal or local interface device.
Storage Areas: A set of six temporary scratchpads in which users may store fragments of text undergoing composition.
Terminal Controls: The ability of user to control margins and page size.
Switches: Special controls needed to regulate the output for those interfacing through microcomputers and intelligent terminals.

TABLE 1 · 1 ***(cont.)***

Evolved Features (Those added to the EIES system based upon feedback from users) *(cont.)*
Reminders: A personalized file of one-line reminders kept by any member which may also be "alarmed" by date and time.
Interests: A file of key words such as "ham radio" which users may enter and associate with so that messages can be sent to all those on the interest list.
Submit & Read: The ability to provide abstracts to others via messages or conference comments which are active keyholes, upon demand, to larger documents stored in notebooks.
Subaccounts: The ability of a group of users to share a single membership slot where only one of the group may be active at any one time.
Games: Various computer games incorporating the ability of players to contribute material to the game or having a communication component (e.g., bridge).
Graphics: The ability to specify simple diagrams through a size independent specification of figures, together with an ability to move windows around in a text item and insert text in windows horizontally or vertically.
Special Programs: Tailored routines for specific purposes. For example, "Terms" collects votes on alternative definitions for tasks such as standards setting. "Respond" administers surveys with multiple choice questions.
Special Communication Interfaces: Tailored communication structures such as TOPICS to deal with Inquires and Responses within a group and allow members to set profiles of their interests for self-filtering of the incoming material.
INTERACT language: A programming language allowing the imposition of special communication or data structures on the basic EIES facility.

*(An * indicates the feature has undergone extensive additions or modifications over the four year operation of the system)*

35 members, and the results are reported in Martino and Bregenzer, 1980.

"SOCIAL NETWORKS"—Group 35 is the "Social Networks Community," which is composed of sociologists, anthropologists, political scientists, and others who share an interest in the study of social networks, or the patterns of "ties" that connect members of a community. As they state in their self description, their "aim is to enhance individual productivity and to facilitate the development of group goals, standards and the like." (N=40; see Freeman and Freeman, 1980).

"SYSTEMS"—Group 40 is "General System Theory." As their principal investigator states, "General System theorists constitute one of the few research communities that are deliberately trying to integrate a wide variety of scientific disciplines. The group plans to use the test facility not only to conduct research, but also to educate each other in the disciplines and approaches involved. As 'common tasks,' the participants will compile a glossary of terms and a 'disciplinary matrix' for the field" (Umpleby, 1977:i). (N=51; See Umpleby, 1980).

"DEVICES"—Group 45 consists of people who share an interest in the development of assistive and adaptive "Devices for the Disabled," and includes disabled persons, research engineers, and consumer-oriented organizations. (N=48; See McCarroll, 1980).

"MENTAL WORKLOAD"—Group 54 is the fifth group which fully participated in the study. "Mental Workload" joined EIES a year after the initial "operational trials" groups. They are concerned with complex man-machine systems, such as the cockpit of a jet plane or the control system in a nuclear power plant. One of their objectives was to publish an "electronic journal." They experienced many difficulties, including the fact that a large portion of their group was British, and the British PT&T (Post, Telephone and Telegraph) would not allow them to use EIES, even though funds had been provided by the British equivalent of NSF. (See Turoff and Hiltz, 1980, for an account of this and other "electronic journals" on EIES. N=21; See Guillaume, 1980 and Sheridan, et. al., 1981).

Also included in the follow-up study is a smaller group (50-INFORMATION SCIENCE) which used the system for about three months. It consisted of about a dozen computer scientists and information analysts interested in the use of systems like EIES for information analysis tasks. An interesting aspect of Group 50 is that they "moved" to EIES from the PLANET conferencing system.

Partial data are available for Group 80, the HEPATITIS Knowledge Base project of the National Library of Medicine. This group included approximately ten medical doctors who are experts on the disease. They used EIES to review and update a synthesis or "knowledge base" of research results related to the diagnosis and treatment of viral hepatitis. They are the only completely task-oriented group for which we have data. (See Siegel, 1980).

In addition to the groups included in this study, many other groups used EIES and some included an evaluation effort which made use of questionnaire items borrowed from the instruments used on the above groups. The data for these groups is not presented in any detail here, since they are not scientific communities, but it is occasionally referred to in the text and may be encountered in other reports on the EIES system:

LEGITECH, a network of state legislative science advisors and resource persons (evaluated by Valarie Lamont).

JEDEC, a standards-setting group of the Electronic Industries Association (evaluated by Peter and Trudy Johnson-Lenz).

WHCLIS (White House Conference on Library and Information Services). The national advisory group used EIES to work with the central staff in Washington to plan the conference (evaluated by Elaine Kerr).

The above groups permitted participant observation in their conferences and activities; these qualitative observations helped to form some of the conclusions and interpretations to be presented.

The study may be considered "quasi-experimental," in the sense that the groups can be compared, and inferences made about the causes of the differences observed. In other words, variations in the characteristics of the user groups as well as among individuals can enable us to search for those factors which explain differences in the observed outcomes. Chapter 3 examines the nature and apparent causes of observed variations among the user groups studied.

CHARACTERISTICS OF THE SUBJECTS

Information from a pre-use questionnaire supplies us with a general picture of the scientists included in this study. For complete percentage distributions on the characteristics summarized below, see the questionnaire in the Appendix.

In terms of employer, 80% of the participants work in academic institutions, 10% are at private research organizations, and only a handful work in business or government organizations. Geographically, the EIES users are spread throughout the United States and Europe, but the largest concentrations are in the Northeast, Middle Atlantic, (including Washington, D.C.), and the West. Only a few are located in Canada or Europe.

Almost all of the subjects are males. Most are between 25 and 44 years old and have a Ph.D. They tend to be "mid career," having received their degrees five to nineteen years previously. A third were in the midst of writing one or more books when they joined EIES, and the majority were working on one or more journal articles. Almost all had published one or more journal articles previously, and about a fifth had published 30 or more articles. Compared to the total population of scientists, then, most of whom have never published anything, the scientists using EIES are considerably more productive than average. They are hard working, with the majority reporting considerably more than the 40 hours which most Americans think of as a "normal" working week. Much of their time is spent teaching, reading professional literature, doing research and writing, with meetings and administrative duties taking considerable time for some.

Although most are not previous users of a computerized communications system, they had used computers and computer terminals before, and had positive attitudes toward computers.

Terminal access was less than ideal. Only about a quarter had their own terminal in their own office. One in five reported no regular terminal access at all. The majority did not have a terminal which they could use at home.

DATA SOURCES AND METHODOLOGY

The most important methodological strengths of this study are that it is longitudinal and employed multiple methods of data collection. Because the data were gathered at several points over a long period of time, we can determine both short and long term reactions to and impacts of the medium. A variety of quantitative and qualitative measures of attitudes and behavior facilitates cross-checks on validity and yields different kinds of insights. The quantitative data enable us to objectively test hypotheses. The qualitative data enable us to gain an in-depth understanding of the experiences of the participants.

Three sources of data were originally planned and have been utilized. In addition, three data sources have been devised using the unique capabilities of EIES. All participants in the study were fully informed about the purpose of the study and the data that would be collected. "Informed consent" agreements were obtained, as required by federal guidelines for the protection of human subjects.

1. Mailed Questionnaires. These took twenty to thirty minutes to complete. They were sent out "pre-use," with a first follow-up at three to six months and a second follow-up at eighteen months. The latter is referred to as the "post-use" questionnaire, even though most participants continued their use of EIES for some time after completing it. Many variables can thus be examined through changes in responses to items repeated on successive questionnaires. The questionnaires and marginals for responses are included in the Appendix.
2. EIES monitor statistics on amount and type of use. These have been obtained monthly, with several cumulative totals incorporated into the questionnaire data file for cross-tabulation.
3. Participant observation. Transcripts have been collected of more than 120 conferences, including some with over 1000 entries. These qualitative observations provided an understanding of what the user groups actually did on the system. The role played might be described as "observer as participant;" the scientists knew that the author was observing for evaluational purposes. As a form of reciprocity, the observer offered to be of assistance whenever possible. A passive role was played, with comments entered by the participant observer generally only in response to a direct request for information or an opinion. Many participants also shared their reactions to the system in private messages to the evaluator, and played the role of informant, describing or calling attention to activities and exchanges on the system which they thought would be of interest.

   In addition, unstructured face-to-face interviews were conducted whenever possible with the principal investigators and/or evaluators of the groups being studied.
4. A routine was adapted to enable EIES to administer and tabulate the responses to short online questionnaires. (This is reported in an article in the Winter 1979 issue of the Public Opinion Quarterly). In addition, the system was used to provide reminders and thank you notes to respondents to the mailed questionnaires. Examples of these are included in the Appendix.
5. Since users are requested to send a message to "Help" (the user consultants on EIES) about any problem they have with the system, a file was created to log all these requests. This file was analyzed every one to three months, and served as a basis for

much of the "formative" evaluation which provided feedback to the designers about system improvements based on user experience. Examples of problems generated through the user consultant file are included in the Appendix.

The user consultants were instructed to remove any material of a personal or private nature and to file only questions and comments received that related to the system. EIES members were informed that their questions to user consultants would be stored in a central file.

6. A file was created of the "who-to-whom" matrix of private messages sent, aggregated weekly.

The resulting data for the first eighteen months can be used to study the evolution of the network of social relationships. The confidentiality of such data is protected by removing the identifying information through a computer routine which substitutes a random number for the "real" user ID. For this study, only preliminary analysis of changes in the size and density of the communication networks over time was done. (See the dissertation by Rice, 1982, for much more detailed analysis of this data file).

METHODOLOGICAL DIFFICULTIES

A long term study always has problems with "mortality" among the respondents. In addition, the needs of evaluators tend to conflict with the priorities of the organization/system being studied. Both of these problems affected this study.

In addition to the usual problem of respondent mortality because of a decline in interest or moving, this study was plagued with very high turnover in EIES membership during the eighteen months. The research design pictured about 150 EIES members beginning use of the system in a particular month, and continuing to use the system for eighteen to twenty-four months. In fact, users straggled onto the system. For instance, some core members of groups 30 and 35 began use of EIES in September, 1977. Groups 35 and 45 were to begin in January, 1978, but problems with delivery of terminals, user materials, paper, etc. meant that many did not actually sign on until the end of February. As a result, it was not possible to find a date when a group could be said to have used the system for three months, which was the originally planned target for the first follow-up questionnaire. The first follow-up was completed sometime between three and six months after the date of first sign-on to EIES.

A more severe problem was turnover of users. Some of this was done informally—a person simply gave their ID to somebody else, and the EIES staff was informed later if at all. Some of it was done purposefully by group leaders, who weeded out inactive members and replaced them with new prospective members. Notification of deletion and replacement frequently did not filter through to the evaluation team and resulted in the absence of a pre-use questionnaire being sent at the proper time. Thus, the number of pre-use questionnaires is the lowest of the three. If a person was deleted from EIES, they might complete a short follow-up questionnaire, but they were not eligible for a post use questionnaire, since they were not on the system long enough. The end result of these problems is that the number of persons for whom we have a complete, three questionnaire longitudinal record is much smaller than the number for whom we have any one questionnaire. A second result of the rotating bodies through the same ID's is that very careful watch had to be made to modify the EIES ID when assigning questionnaire ID's. For instance, assuming there was more than one person with the ID of 300, then the first one was labelled 1300, the second 2300, the third 3300, etc. There may be a few errors where the "wrong" person's questionnaires and monitor data are being matched, although we spent a great deal of time trying to clean the data of such errors. A related problem is that "3–6 months" and "18 months" are very rough descriptors for the time of data collection. A few respondents had been active somewhat longer than the target figures at the time they completed a questionnaire, and many for a shorter period of time.

One of the most severe methodological problems is the danger which anthropologists refer to as "going native." In order to understand the use of EIES and its evolving electronically based social system, and to remain in communication with the subjects, it was necessary to spend a great deal of time online. More than 3000 hours online have been logged in the course of this study. Thus, the objectivity afforded by "outsider" status was long ago lost.

The main solution to the "going native" problem is that the data presented and interpretations made stay as closely as possible to objective evidence supplied by the participants themselves—monitor data on the amount of use, questionnaire responses, excerpts from conferences and messages on EIES. In other words, this report tries to summarize what the objective data say, and to minimize acquired biases of the participant observer.

METHODOLOGICAL WEAKNESSES OF THE EIES FIELD TRIALS

The EIES project was a unique approach to studying factors relating to the organization and productivity of scientific specialties: actually changing the communication modes of several specialties, and then figuratively sitting inside the communications network to observe what happens. However,this field experiment cannot measure all that might occur should CC become a "normal" widespread, nonexperimental mode of communication.

1. A New Technology Limited to a Few Groups. One analogy that might be made is to the situation when telephones were new and owned by only a few persons. Just as people used to have to shout to be heard over long distance and much static was commonplace, a few technological kinks in the system, which may discourage and frustrate users, can be expected in the beginning.

Secondly, the scientist-users had to resort to other communication modes for other roles they play. Eventually, terminals in the home and the use of computerized conferencing might become as cheap and widespread as TV ownership is at present. At that point, people could belong to many "conferences," corresponding to all their roles—a family news conference, for example, and a chess conference. For the field trials, however, only the approximately 300 scientists on the system could reach out by computerized conferencing. As a result, use of the system was added on to use of other communication modes rather than replacing much of their use. A related factor is that for system planning purposes, the specialty group's ability to expand to include new members on the system was arbitrarily limited. If CC were a generally available service like the telephone, any number of additional persons might join the network. Still another factor related to the newness and scarcity of the technology is that many of the scientists never before used a computer terminal and might not have had any other use for it; thus, the investment of time to learn the system might be problematical. Since users did not generally have a terminal both at home and in the office, they had to take the trouble to carry it around if it was to be available at all times. If the day ever comes when terminals are as omnipresent as TV's, they will always be conveniently at hand without foreplanning, and used with as much frequency and ease as more familiar household appliances are now.

2. Unknown Generalizability. To what extent are the findings reported here limited to the EIES system and scientific users? To what extent might they be an artifact of the acquired biases of the evaluator? One way to deal with the question of generalizability is to examine comparable data collected by other observers for other systems and types of users. Whenever possible, the data collected for this study were coordinated with those being collected for similar studies, so that comparisons could be made. The main comparable studies are summarized below. A later chapter directly compares the results of questions that were asked in two or more studies.

OTHER STUDIES WITH COMPARABLE DATA

"Theory Net," a comparable "invisible college" in the area of theoretical computer science, which used the TELEMAIL computer-mediated communication system, was studied using some of the same measures as were employed for the scientific communities on EIES. Some of the measures included in this study of EIES are replications of indicators used by Robert Johansen and his colleagues at the Institute for the Future in their studies of PLANET users. Some replicate measures used by Edwards in her study of NLS (now called AUGMENT). In addition, studies of the Swedish COM system and of the "Terminals for Managers" program at Stanford include similar questions and indicators.

PLANET CONFERENCING SYSTEM DATA

Approximately 500 members of more than 18 organizations were observed using PLANET (or, in a few cases, the related FORUM system), by Vallee, Johansen, and their colleagues as part of a project conducted by the Institute for the Future. PLANET is a simple conferencing system with limited message capabilities and no "notebooks" (personal files and text processing capabilities) or associated data bases. Among the organizations were NASA, the U.S. Geological Survey, ERDA, and the Kettering Foundation. The conferences lasted from one week to twenty-four months. Applications included topical conferences on food and climate, individually guided education, technology transfer, and psychic research, as well as the management and coordination of technical projects or joint report writing (Vallee et al., 1978:xv). 188 of the participants responded to a post-use questionnaire. These tended to be the heavier users of the system, with 40% of the respondents above the highest quintile in terms of number of sessions and another 30% between the highest and the second quintile (Vallee et al., 1978:109). We thus have a very wide range of sizes and types of groups and applications, plus an unrepresentative set of survey responses. Nonetheless, comparative responses for the same post-use questions included in the EIES and MACC TELEMAIL field trials may be informative.

EDWARDS' STUDY OF NLS

NLS (oN Line System, now called AUGMENT) is a general office support system. Particularly when used in conjunction with an intelligent terminal with a special "mouse" device for pointing during editing, it is excellent for document production. It also includes three communications capabilities: asynchronous message exchange, real time messaging, and file exchange. It does not include a conferencing component.

Edwards' (1977) study was based on a questionnaire sent to 250 users of NLS in thirteen organizations. Ninety-four, or 38%, responded. Of these, 30% were managers, 42% researchers, and 28% support staff. Some of the researchers also had a supervisory role, as a total of 40% reported some supervisory responsibility.

The NLS setting was quite different from the EIES setting during the operational trials. It was used as a tool to directly support the regular, paid job. It is therefore most important in increasing the generalizability of the EIES findings that many of Edwards' findings about the importance of attitudinal variables are similar. A copy of Edwards' questionnaire was made available during the design phase of our study. Many of the items were borrowed to increase the comparibility of the findings.

THE TFM MESSAGE SYSTEM

Rice and Case (1981) studied the impacts of a computer-based massage system called TFM (Terminals for Managers) on 89 senior-level university managers at Stanford. A "Time 1" questionnaire was administered within ten weeks of their introduction to the system, and a "Time 2" questionnaire some two to five months later. Some of the questions they explore parallel those in the EIES case study, and their results will be compared whenever appropriate.

THE COM CONFERENCING SYSTEM

The final related study which will be referred to frequently is the COM computerized conferencing system in Sweden (Palme, 1981). COM (and its Swedish laguage version, KOM) has many features like those of EIES and PLANET. For instance, it has "open conferences" and "closed conferences" (like public and private conferences on EIES), messages, search and retrieval capabilities, text editing facilities, and voting facilities. A unique aspect of this system is that COM was installed on five different computers within Sweden plus some outside Sweden, with the capabilitiy for automatic transferring of items among the different computers. Such a decentralized design is important for saving "line time" in international communications. The COM evaluation included motitor statistics, observation, and questionnaires and interviews administered by Adriansson (1980).

THE THEORY NET GROUP ON MACC-TELEMAIL

"Theory Net" was a group of theoretical computer scientists using an electronic mail system for much the same purposes as those for which the various "invisible colleges" of scientists used EIES. This was a small group, including initially only twelve individuals at nine institutions. They were sent versions of the same pre-use and follow-up questionnaires as were administered to the EIES users.

Self Reported Characteristics of Theoretical Computer Science as a Specialty. The scientists in the Theory Net group are rather young; most are under 35. All have Ph.D's. They do not write books in this highly mathematical field, but they were working on an average of five journal articles and almost all published one or more articles the preceding year. They had spent most or all of their scientific careers in the specialty. Most consider themselves to be in the middle to higher range in terms of professional reputation within the specialty. As computer scientists, all were of course very experienced in the use of computers and terminals before using the TELEMAIL system, and had favorable attitudes toward computers. However, on the basis of their previous experiences, they tended not to trust computers for the storage of paperwork used daily. They anticipated using the system only for private messages and reported that they were strongly motivated to use the system.

There was unanimity that there is an "intellectual mainstream" in this area, and all of the participants felt they were "in" the mainstream. Competition is generally perceived as moderate and mainly attributed to the high achievement drive of some of the members of the specialty area and to competition for funds. There were no reports of strongly opposing theoretical viewpoints or of unethical competition.

The picture which emerges is thus of a somewhat more mature specialty area than was typical of the research communities on EIES (see Chapter 3). This is reinforced by the reported preference for working in "established areas."

The MACC-TELEMAIL System. The TELEMAIL system (later named @MAIL, when TELENET took over the right to use the name TELEMAIL) provides the ability to send items such as memos, drafts of working papers, and computer program source listings or data to other "addresses" or "mailboxes." It is resident on the Univac computer at the University of Wisconsin, Madison, and accessible through the TELENET network. Its design was influenced by other computer-based message systems, particularly HERMES. It has a simple set of commands that suffice for the beginner:

EXPLAIN
STATUS
PRINT
TO
MAIL
DELETE
EDIT
QUIT

There are also more complex features, such as the file system, "filters," and a separate EDIT system. (Academic Computer Center, The University of Wisconsin, Madison, 1975, 1977. Updated manuals are now available. These are the versions originally supplied to Theory Net members.) The "mail" metaphor pervades the system, with "postmarked" dates and even a "Post=Master," the "mailbox" to which questions can be sent. Note that in order to edit, a user had to enter a separate edit system, and when the editing was finished, re-enter the MAIL system.

The user interface includes conventions peculiar to UNIVAC, with the use of asterisks, periods, and such to name subfiles. For instance, a sample command is:

COPY SOURCECODE*FORA.PROGRAM to JIMMY=CARTER

Such a copy command had to be used after an edit, before a message could be sent. As a result, many Theory Net members avoided the editor.

Methodological Difficulties. The Theory Net Group communicated almost entirely by private messages. It was therefore not possible to observe their behavior or to become accepted as a neutral and sometimes helpful observer, as in EIES. An attempt was made to gain rapport by setting up a group file, which could act like a conference. There, a plea was made for copies of the material being sent among members of the group, so that it could be analyzed. One person cooperated by sending some sample messages; everyone else ignored it.

Questionnaire data are sparse and incomplete. This group started very small, and was frequently added to. Unlike the arrangement with the EIES staff, a copy of the pre-use questionnaire was not automatically sent to each member as he/she was added. We therefore have very incomplete "pre-use" data; it includes only the original core group of members. There were eight responses to the pre-use questionnaire, which was sent out in the early fall of 1978. There was no obvious point at which to send follow-up questionnaires; at about the three to six month point, when they had been sent for EIES, there were plans for the Theory Net group membership to be greatly enlarged, and it thus seemed premature to do a follow-up. Thus the follow-up was administered after approximately eighteen months (equivalent to the post-use questionnaires for EIES), and no comparable post-use measure was taken. There were twenty-two follow-up responses from the expanded Theory Net group.

In looking at comparable data for Theory Net on MACC-TELEMAIL and for the EIES groups, any differences or similarities observed can be interpreted as supportive of hypotheses, but not as proving or disproving hypotheses. There are too many differences in the nature of the subjects studied, the systems used, and the timing of the data collection, plus poor response for the Theory Net group, to rule out many alternative explanations for any similarities or differences. No statistical tests of differences between the two sets of data will be made, since the data themselves are not fully comparable. However, the data can be useful in helping to identify some "universals" that seem to hold accross very different types of computer-mediated communication systems. (For a full account of the Theory Net study and findings, see Hiltz, 1981, Chapter 8).

SUMMARY

This case study is an in-depth, long term analysis of the use of one particular computer-mediated communication system by scientific research communities. Its generalizability is increased by a number of comparable studies of similar systems by other types of "knowledge workers."

The use of the computer to facilitate communication and "knowledge work" by processing text as well as numbers is a recent technological innovation which is also a "social invention" in Whyte's (1980) terms. Linked through computer networks that may be international in scope, dispersed persons may exchange communications and work together despite geographic separation and differences in time zones. Whyte's (1982) description of the methodology and goal of evaluation of a social invention serves as a good description of the procedure followed in this study:

> Having discovered a social invention, you then move into observe, interview, and gather documentary materials so that you will eventually be able to provide a systematic description of that invention. You then seek to evaluate the invention . . . This is not simply a matter of judging the degree of success or failure. If the invention appears to work, this judgment does not tell us *WHY* or *HOW* it works. If we are to be able to describe a social invention in a way that makes it potentially useful in other situations, we must grasp the social principles underlying its effectiveness . . .

A technological innovation is merely a potential, which may be accepted or rejected, and used effectively or ineffectively, depending upon the mediating influence of a variety of social forces. This mediating social context includes the preexisting skills, basic personality attributes and values of individuals who are exposed to it; and the preexisting social structure of the target groups for the innovation. This study emphasizes the key role played by social processes in determining the acceptance and impact of a computer-mediated communication system called EIES. The findings have implications for the successful implementation of "office automation" technologies in the future.

The title of this book is "Online Communities." It was chosen to suggest the idea of a new kind of professional and technical community: one that is defined not by working together in the same physical office space, but by common membership in a computer communications network.

All social systems are characterized by the same fundamental processes, including division of labor, status differentiations, shared norms and expectations, and patterns of friendship or alliance. Perhaps computer-mediated communication networks can best be thought of as a "new kind" of social system: one in which the familiar social processes in the workplace and the organization become subtly altered by the new mode of communication.

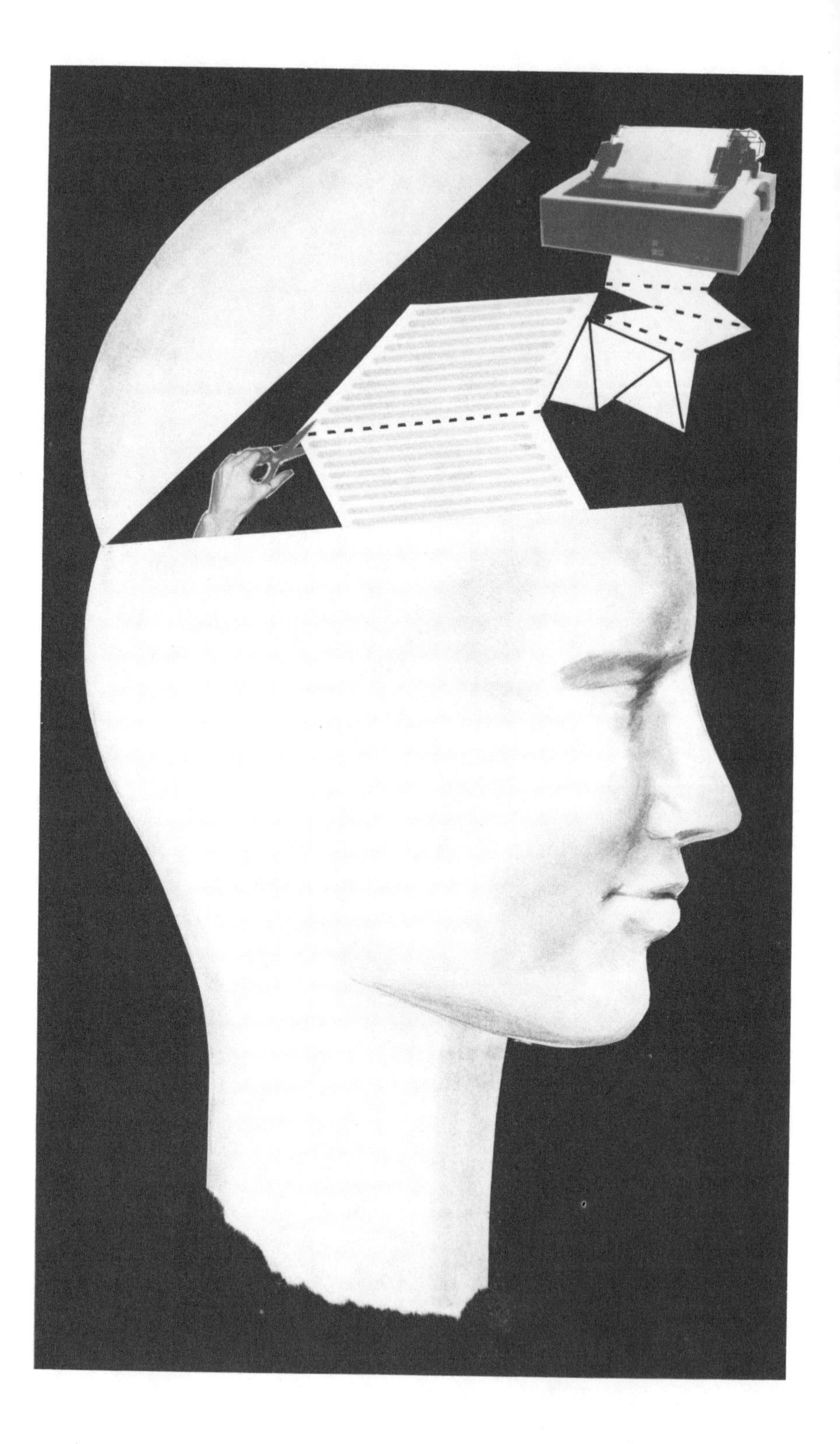

2.

DETERMINANTS OF USE OF THE EIES SYSTEM

One of the most intriguing aspects of computer-mediated communication systems is the contrast between users who integrate this new form of communication and information exchange into their lives and those who do not use it at all, even if they have free access. What explains or predicts acceptance of a system such as EIES?

Systems designers often assume that if they produce a "good" set of programs and if they are adequately documented, they will be used. However, a new computer system, particularly one meant to augment the work of professionals such as scientists, must be looked at as an innovation which may or may not be accepted within the social context in which it is introduced. Factors which have little to do with how well or poorly the system is designed may determine its acceptance or rejection by the target users.

In this chapter, we will look at which variables explain differences in amount of EIES use. (The other aspect of user acceptance, subjective opinions of the system, will be examined in a later chapter). The data point to one overall conclusion: it is motivations of the participants and their location within a particular social context, not system characteristics, which are the primary determinants of use, at least in terms of initial system acceptance. Users are sensitive to specific system characteristics; they influence subjective satisfaction, the choice among available systems, and the range of professional activities for which a computer-mediated system will be used. But variations in system characteristics seem to be of secondary importance, as long as minimal criteria of reliability and ease of use are met.

TABLE 2 · 1 ***Characteristics of Individuals Which May Affect System Acceptance***

A. Attitudinal variables
1. Attitudes toward task
   a) Relative importance or priority*
   b) Degree of liking or disliking of the task (pleasant/unpleasant, challenging/boring, etc.)
2. Attitudes toward media
   a) Attitudes towards computers in general*
   b) Expectations about the specific system
      1) Anticipated usefulness (amount of use)*
      2) Anticipated impacts on productivity*
      3) Anticipated difficulty of use
   c) Attitudes toward alternative media (telephones, writing letters, travel, etc.)
3. Attitudes toward the group (liking, respect, whether they are an important reference group)
4. Expectations about how system use will affect relationships with the group*
5. Perceived pressure to use the system*

B. Work-Related Skills and Characteristics
1. Personal communication skills
   a) Reading speed*
   b) Typing speed*
   c) Preference for speaking or writing*
   d) General literacy (writing ability)
2. Previous related experience
   a) Experience using computers*
   b) Use of computer terminals*
   c) Use of other computer based communication systems*
3. Physical or intellectual disabilities
4. Productivity
   a) Hours per week worked*
   b) Number of publications or other output measures*
5. Connectivity
   a) Number of persons in field with whom one is in contact*
   b) Number of persons on system with whom one was in previous contact*
   c) How well known person is in field*
   d) Whether a scientist feels "in the mainstream" or not*
   e) Number of coauthors (or coworkers)*

TABLE 2 · 1 ***(cont.)***

C. Demographic characteristics
1. Age*
2. Sex*
3. Educational level*
4. Race, nationality or subculture

D. Environmental variables
1. Available resources, including secretarial support
2. Position in the organization (or status in informal group)*
3. Amount of pressure to use the system (from superiors and peers)*

E. Psychological variables

* *Indicates that one or more measures of this factor were included in this study.*

CONCEPTUAL FRAMEWORK

The determinants of acceptance and use of computer-mediated communication systems as listed by Vallee et al. (1974:22) can be expanded and grouped into these factors: INDIVIDUAL USER, SOCIAL GROUP OR ORGANIZATIONAL CONTEXT, TASK, the SYSTEM itself, and the EQUIPMENT which interfaces with the system.[1] These factors may be treated as competing hypotheses or alternative explanations for predicting system usage.

The detailed lists of potentially relevant characteristics of the individual and of the social group are shown in Tables 2.1 and 2.2. Within the context of studying only five EIES groups (which did not have any particular task and which were confined to a single system, with little variability in available equipment), most of the attributes of Task, System and Equipment that have been developed could not be included in this study of determinants of system use.

In regard to the SYSTEM and EQUIPMENT, we have some data on the effects of:

DOCUMENTATION (was it clear and comprehensive, or not?)
Whether or not there was a "live" teacher provided

[1]This framework was expanded and developed in a workshop project funded by the Division of Information Science and Technology, NSF. Contributions were made by Murray Turoff, Valarie Lamont, Elliot Siegel, and John Senders, as well as the author of this report, who was simultaneously P.I. for the workshop project.

TABLE 2 · 2 ***Group Factors Which May Affect System Use***

A. STRUCTURE
   1. Size*
   2. Degree of geographic dispersion
   3. Centralized vs. decentralized control
   4. Pre-existing communications ties or network
B. LEADERSHIP
   1. Style
   2. Level of effort or activity by the leader*
C. COHESIVENESS
   1. Socio-metric ties
      a) Have they met face to face?
      b) How many members of the group are known to each other before they begin communicating on the system?*
      c) Have they worked together previously?
      d) Do they form cliques, have many "individualists," or are they an integrated group?*
   2. Competitiveness*
   3. Trust or openness among members*
   4. Status (are most group members prestigous in their fields, or not?)*

** Indicates that a measure of this factor was included in this study.*

Quality of the TELENET interface (whether or not TELENET was a source of "trouble")
Whether or not the system was a source of difficulties
System Availability (downtime during workday or unavailability nights and weekends)
Trouble with the telephone or high cost of long distance telephone due to absence of TELENET node
Access to terminal (own or share or none at office; own or available loan or none at home)
CRT, print, or both
Size and weight and printing speed of the terminal(s) available.

PROCEDURE

We have two sources of independent variables in exploring the determinants of the amount of EIES use. The first consists of data from the follow-up questionnaire in which the respondents report what factors are important in limiting their use of EIES. The problem with such subjective reports is that users may be unwilling or unable to report the "real" explanations for low levels of use. We can check the accuracy of such perceptions by seeing if the reasons given as "very important" in limiting EIES use are reported more often by low level users than by high level users.

The second type of data are observed correlations between reported pre-use attributes of individuals and user groups, and their observed levels of system use. Such evidence is subject to the problem that "correlation does not necessarily mean causation"; an observed correlation may simply mean that two variables are both affected by some third variable, rather than that one causes the other. In order to sort out the patterns of causation, we do have the advantage of time-ordered data, and we can also use multivariate analysis to help determine the patterns of relationship among the variables that we are examining.

Correlation and significance statistics will be used to categorize observed relationships as strong, moderate, weak, or non-existent.[2]

MEASURING LEVEL OF EIES USE

An overall profile of the "average" (mean) use of EIES during the operational trials is shown in Table 2.3. This is derived from monitor data on the cumulative activity of all EIES members as of April 1, 1980. At that time, more than half the participants were non-scientific users, and the operational trials groups had been deleted and are not included in

[2]When examining statistical association, the most frequent measure will be gamma, which is appropriate for linearly related ordinal variables. Occasionally, the pattern of correlation is curvilinear, in which cases we will report eta, a measure of curvilinear correlation. If the dependent variable is at an interval or ratio level of measurement, Pearson's R is used.

Chi square tests are used for all cross tabulations to estimate the statistical significance of the patterns of association. The results of the chi square tests (p, or probability level) should be interpreted as a very rough measure of the extent to which the number of observations and the patterns of association observed are large enough to serve as the basis for conclusions. A "p" (probability) of less than .05 is the traditional level for statistical significance. Since the respondents do not represent a random sample of all users of EIES, let alone of all potential users of all such systems, chi square or t-test results cannot be interpreted in terms of a level of confidence in generalizing to such a larger population.

In looking at correlations of pre-use attitudes and characteristics with subsequent hours online, we will refer to correlations of .10 to .20, accompanied by probability levels of .10 or less, as being "weak" relationships. If the correlation is less than .10 or the significance tests indicate that the probability that the results could be accounted for by sampling error or chance is greater than .20, we will say that there is "no relationship." Moderate relationships refer to correlations between .20 and .49, with at least a .10 level for significance. "Strong" relationships will be said to exist for those that are characterized by correlation coefficients of .50 or greater, significant at the .05 level or better.

In looking at directly reported reasons explaining limited use of the system, we will call those named by 20% or more "strong"; 10–19% "moderate"; 5–9% "weak"; and less than 5%, not a determinant.

TABLE 2 · 3 *Average User Profile*

Category	Amount
Hours Used	105.5
Number of Sessions	265
Average Session Time (minutes)	24
Text Items Composed	279
Text Items Received	1,194
Items Transacted/Session	5.6
Average Input Rate (words/minute)	7.9

Subsystem	% of Items Composed	% of Items Received	Size (Lines)	Circulation Ratio
Messages	69.1	35.8	10	2.2
Conferences	22.3	60.9	14	11.7
Notebooks	8.6	3.3	19	1.6

Source: Accumulated Monitor Statistics as of April 1, 1980

the computation of the average. The data do give us a rough idea of the usage patterns of participants. For instance, we see that users did most of their sending in the form of private messages, which go to about two persons on the average; but most of their reading is in the conferences, where items are read by about twelve persons, on the average. We also note a fairly long average session length (24 minutes).

However, usage is highly skewed. Table 2.4 shows the surprising fact that 40% of the scientists invited to have free access to EIES either never signed on at all, or dropped out before learning to use the system. Within this "dropout" category, 11% of the sample never signed on at all.

In a system such as EIES, when use is voluntary for most members (such as during the operational trials), amount of use in terms of hours online can be taken as a fairly valid measure of user acceptance. However,

TABLE 2 · 4 ***Hours Online At Follow-up, by Group***

Group	Group #	<5*	5–19	20–49	50+
30(Futures)	(N=35)	34%	20%	29%	17%
35(Social Networks)	(N=40)	32%	25%	28%	15%
40(General Systems)	(N=51)	33%	33%	22%	12%
45(Devices)	(N=48)	58%	25%	13%	4%
50(Information Sci)	(N=8)	12%	62%	13%	12%
54(Mental Workload)	(N=21)	62%	29%	5%	5%
80(Hepatitis)	(N=10)	20%	50%	20%	10%
Total	(N=213)	40%	29%	20%	11%

**Includes persons who never signed on*
Source: Monitor statistics for cumulative time online, June 1, 1978, or beginning of month when follow-up was returned

lack of use in the totally "voluntary," almost "extra-curricular" mode that characterized the operational trials cannot be assumed to validly indicate rejection of the system. It simply indicates that the relative costs and benefits were more favorable for off-line activities. (In other words, low use has to be accompanied by poor opinions of the system in order to indicate active "rejection" of the system.)

Since use was skewed and our independent variables are mostly nominal or ordinal, cumulative hours online has been divided into levels or categories for most analyses. This procedure has the advantage of not weighting the small number of users with very high numbers of hours of use too heavily in the analysis, and of being straightforward and easy to understand. For these data, we found that it has much the same analytical effect as using the log of the number of hours, in those analyses where both methods of handling the dependent variable were tried and compared.

The first level consists of those who did not accept the system. They never signed on at all, or did not stay online long enough to get through the learning period and be able to use the system effectively. (This is less than five hours total use, referred to as the "dropouts"). "Low" use level is 5 to 19 hours; "intermediate," 20 to 49 hours; "high" use 50 to

99 hours online; and "very high" is more than 100 hours of connect time. These break points correspond to observed changes in user behavior derived from monitor and questionnaire data, as well as giving us reasonable marginal distributions among the levels.

These data are available for cumulative hours online at follow-up, post use, and several other points in time. The follow-up data have been chosen as the focus for this analysis. One reason is that this is the point for which we have the most questionnaire data. Even the "dropouts" were sent a two page follow-up, asking for a ranking of reasons for not using the system. Responses to the short (for dropouts) and long follow-up questionnaires totalled 195 out of 213, almost twice as much questionnaire data as are available if the post-use questionnaire were used. Another reason is that "acceptance" or "rejection" can be fairly clearly established in the first three to six months. If a person does not use the system in that time, they are very unlikely to ever use it. In fact, many of the non-users were subsequently dropped from the system by the group leaders.

GROUP VARIATIONS

Useage patterns varied markedly among the scientific communities (see Table 2.4). Group 54 (Mental Workload) had the highest dropout rate (62%). Many of these were the British users who were refused access by the British Post Office, whose actions and their impact have been described as follows (Sheridan, et al., 1981:21)

> Following the original approach to the British Library . . . the B.L. drew up a contract which would allow up to 12 U.K. scientists to participate in the "Mental Workload" electronic journal. The B.L. approached the U.K. Post Office External Telecommunications division in late 1977, and by the beginning of 1978 received permission to use Telenet to carry out the experiment internationally. Unfortunately, although the Post Office received all the documents, they did not reply in writing to the B.L. In early December, 1978, after several U.K. participants had bought terminals, rented recorders, etc., . . the Post Office suddenly announced that they had not realized that "messages" would be transmitted, that P.O.–Western Union had an agreed international monopoly on transatlantic communication, and that no one, under any circumstances, could use any medium other than the P.O.–Western Union system. Despite efforts by the B.L., they flatly refused to help in any way or discuss the question, make an exception for an experiment, or do anything . . .

This was very demoralizing to the group. Together with lack of terminal access for some U.S. members and the feeling by others that the actual work of the group as it evolved did not converge with their interests, the projected mental workload group of 42 participants was reduced to only 21.(Sheridan, et al., 1981: 22).

Group 45 (Devices) also had a large number of invited participants who never became active. The lowest dropout rates were among the two task-oriented groups (50,Information Science, and 80, Hepatitis). These also happened to be the smallest groups; thus, if there is any overall relationship between group size and amount of use of a system by its members, it cannot be determined from the operational trials groups. We will look at a few group-related variables which seem to predict amount of use of EIES in this chapter, related to the perceptions of the members about the competitiveness or unethical behavior of the members and total self-perceived status level of the group's members. Other group factors which may explain these variations are further explored in Chapter 3.

SUBJECTIVE REPORTS OF FACTORS WHICH LIMIT USE OF EIES

Table 2.5 shows the importance ratings for factors which limited use of EIES, at follow-up. These responses lump together the dropouts, the very heavy users and all those in between. An additional set of tables broke down the responses by level of use. Surprisingly, there are not many differences by level: reasons given as very important by those who never used the system or used it very little are almost the same in terms of frequency of mentions as those given by heavier users. The main results of these cross tabulations can be discerned from the correlation coefficients reported in Table 2.5, in conjunction with the results of the Chi-square test which indicates the level of statistical significance of the observed correlation. A minus sign in front of the correlation coefficient means that the reason was given more frequently by dropouts and low level users than by high level users.

The reasons in Table 2.5 have been listed in order of the frequency with which they were named as "very important" by all users, with some weight given to the frequency of "somewhat important" responses. We see that conflicting demands and priorities are by far the most frequently reported barrier to use. Overall, 47% of users report that an important limitation on their use is that "other professional activities must take higher priority." The frequency with which this reason is indicated is somewhat higher for the dropouts and infrequent users, indicated by a Gamma of $-.17$. This weak relationship with hours online is not statistically significant ($p=.16$).

Qualitative data from the post-use questionnaire reinforces the im-

TABLE 2 · 5 ***Importance of Reasons Limiting Use of EIES, And Correlation (Gamma) with Level of Use***

Reason	Very Important	Somewhat Important	Not Important	Gamma	P
Other professional activities must take higher priority	47%	30	22	−.17	.16
Limited night or evening hours	20%	21	60	.27	.05
Inconvenient access to a terminal	19%	18	63	−.15	.16
Trouble with Telenet	15%	19	65	.31	.01
Had some bad experiences	11%	31	58	.29	.005
The system is too complicated	9%	25	66	.12	.001
Trouble with telephone	10%	17	74	.20	.05
Cost of telephone or Telenet	9%	11	80	.08	.45
There is no one on this system with whom I wish to communicate a great deal	7%	16	77	−.40	.14
The conference comments or messages I have received do not seem worth reading	7%	31	62	−.01	.05
Red notebook documentation looked like too much to read	6%	27	68	−.13	.04
Inadequate leadership of the group	5%	17	78	.14	.59

TABLE 2 · 5 ***(cont.)***

Reason	Very Important	Somewhat Important	Not Important	Gamma	P
I am not very interested in the subjects being discussed	6%	17	77	−.02	.60
I do not know how to type	5%	15	80	.08	.54
I do not like using a computer system like this	3%	8	89	.15	.76

Source: Follow-up questionnaires sent to Groups 30, 35, 40, 45, 50, 54, 80.Total N responding is 195
Note: Gamma = correlation with accummulated hours online at follow-up, categorzied by level. A "minus" gamma indicates that the less time online, the more likely the person was to name the reason as very important.
"p" = probability that the correlation could be attributed to sampling error, based on Chi square test. A p of .05 or less is generally considered "statistically significant".

portance of the relative priority of the task in determining level of use of the system. Many respondents indicate in their open-ended comments that the work for which they are being paid conflicts with use of EIES. In fact, many see EIES as taking away from the time needed to do their official job. Communication with ones peers in other institutions is simply not as high a priority as the work commitments pressing in at the workplace. A selection of such comments, from the open-ended question on the post-use questionnaire, is shown in Table 2.6.

A related motivational variable is having "no one on the system with whom one wishes to communicate a great deal." Though only 7% of all EIES respondents list this anti-social sounding reason as "very important," those who do feel this way are likely to be dropouts. Not a single user in the sample who did not particularly want to communicate with the limited community online logged over 50 hours, and the correlation (gamma = −.40) is the strongest for any of the self-reported reasons for non-use.

After the motivational variables of conflicting priorities and lack of desired communication partners, but far behind, are factors that have to do with system access. "Limited night or evening hours" was a strong enough deterrent so that steps were taken to put EIES up seven days a week, around the clock. During nights and weekends, someone is not always at the console in case of a crash, but a system was devised whereby EIES can be restarted remotely, by telephone, if it is found to have crashed.

TABLE 2 · 6 ***Explanations Offered for Low EIES Use***

Post-Use Open Responses Emphasizing Priority Conflicts

1. I'm very busy, with heavy commitments. EIES doesn't contribute to any of the things I really MUST do. It is a peripheral interest."
2. Too busy with other things.
3. Time pressures resulting from need to EARN by consulting and teaching extra loads.
4. Lack of time—other research projects are more pressing.
5. Lack of time and pressure of my business—I am associated with a small R&D firm which implies a constant need to seek new contracts.
6. I am under a great deal of time pressure.
7. I work full time and am a full time graduate student and half time mother—need I say more?
8. Other matters, with more immediate DEADLINES, kept interfering.
9. Very busy with other things such as classroom teaching; talking with students; working on articles and proposals; committee work.
10. External pressures for time keep me elsewhere. Except for a few direct research collaborations over EIES, the rest seems more like an interesting luxury than a necessity.
11. There is no job related reward. EIES takes time and is not recognized by the university . . . this is unfortunate.
12. Pressure of administrative responsibilities.
13. It is extremely difficult to match full time (university) professional interest and responsibilities with those generated by the wide membership of EIES.
14. Extremely busy schedule during last year.
15. Lack of time to participate. THIS IS THE ONLY reason.
16. Work pressure.
17. Other time consuming work is more pressing.

Question: "What one or two factors best explain why you have not used EIES more?"

Another access barrier ranking high on the list of factors which decrease use of EIES is trouble with the TELENET link. (The more time they spend online, the more trouble they have. And TELENET's reliability has been decreasing, not increasing. See the discussion below). Closely behind this is the related access barrier of trouble with the telephone connection. But note that reporting of all of these access barriers *increases* with use. . . . in other words, encountering access difficulties does not cause low use, but is rather proportional to the amount of use.

The one frequently mentioned access barrier which does appear to be a cause of low use is inconvenient access to a terminal.

Characteristics of the system—having bad experiences such as a crash, or the feeling that it is "too complicated"—are "somewhat" important reasons cutting down use, but are not very important to many users. "Bad experiences" peaks in the low use range (5–19 hours), where 40% say this has been "somewhat" important in cutting down use.

The relatively low prominence given to cost is probably attributable to the subsidized memberships of the users. They generally had to pay only local telephone charges to reach a TELENET node. For non-subsidized users, cost would undoubtedly be a more important factor accounting for level of use.

TELENET TROUBLES

TELENET was annoying enough to users that they were motivated to establish and actively participate in a public conference to air "TELENET Experiences." Begun at the end of the operational trials, it acquired 72 entries in the first month, most of which are descriptions of difficulties. The number of TELENET difficulties encountered during a month by all users is undoubtedly many times that which users take the time and trouble to document in the public conference. Summary of this transcript indicates just what sorts of TELENET problems users frequently encounter.

1. Local TELENET nodes become overloaded; they simply do not answer when dialed or they give a busy signal.
2. One or more local nodes goes out of service. If it is the Newark node, then no one can reach EIES through TELENET.
3. Users are dropped by TELENET and are "frozen" online. More specifically, somewhere in the network, the fact that the user is connected to EIES gets lost. The packet loses its address, so to speak, and does not get delivered to the EIES computer. The user inputs and gets no response, because EIES receives nothing to respond to. Meanwhile, the port on EIES sits open, with EIES

waiting for the lost packets that never arrive. If the user hangs up and redials, she or he discovers that, "That ID is in use." EIES has received no signal that the user hung up the phone, and keeps the line open until either the automatic time out occurs (for which the default is set at twenty minutes) or a privileged EIES staff member "bumps" the frozen ID. Users find themselves, if they know someone else's access code, in the absurd position of signing on as someone else in order to request that their own ID be "bumped." Or they call Newark. Or they impatiently wait for twenty minutes. Or they give up and end the session.
4. Most seriously of all, TELENET sometimes mixes up packets and switches users, even among different computer systems.

Along with what most users felt was a constant decline in the quality of TELENET services, the fall of 1980 brought a rise in price for TELENET: from $3.75 to $5.00 per hour—a sufficiently large increase to constitute an economic problem for many users supporting their own account charges. As one user summed up the situation (Douglas Cayne, in Conference 1011, the public conference on Telenet Problems, comment 35):

> If the networks can do no better than offering this sort of consistently poor—borderline unusable—service, it may be many more years than we've been predicting before we become the Network Nation, or before people find computers useful enough to have in the home . . .

REASONS GIVEN BY DROPOUTS

A subgroup of particular interest is the "dropouts" (those who accumulated less than five hours of line time). The following are the only reasons listed as "very important" by 10% or more of dropouts:

Other professional activities	55%
Terminal access	19%
Limited night or weekend hours	12%
No one to communicate with	12%
Trouble with telephone	11%
Material not worth reading	10%

Looking only at the reason named as the single "most important," conflict in priorities with other professional activities is the only reason given with great frequency by the "dropouts" (those who never spent more than five hours on EIES). The second most frequently listed "most important factor" by the dropouts is inconvenient access to a terminal, named by 9%. (The complete table of these data is not included here. Almost all reasons, except the above two, are named as "most important" by only a small number of people).

PREDICTORS FROM THE PRE-USE QUESTIONNAIRE

Many of the pre-use questions, measuring motivation to use the system before having any experience with it, turn out to be significantly correlated with subsequent amount of use. This includes anticipated value of the system (Gamma = .27) and amount of time spent on the pre-use questionnaire (Gamma = .30). The latter may seem to be a surprising predictor, but it is an interesting behavioral measure of pre-use attitude toward the system and the project. The strongest predictor is the amount of time which a prospective user estimates that s/he will spend online each week (Table 2.7). Two thirds of those who felt that they would spend less than 30 minutes a week online became dropouts.

On the other hand, most of the "objective" characteristics of users that might be thought to predict acceptance, such as typing speed, did not turn out to be related to amount of use.

Estimated number of sign-ons per week, before the system was used, follows the same pattern as anticipated time online per week. A third

TABLE 2 · 7 ***Anticipated Weekly Usage of EIES, Before Use, by Time Online at Follow-up***

	< than 30 min.	30–60min.	1–3 hours	4 hours
< than 5 hours	62%	35%	40%	4%
5–19 hours	25%	50%	20%	20%
20–49 hours	13%	15%	29%	44%
50+ hours	0%	0%	11%	32%
N	8	20	35	25
	100%	100%	100%	100%

Chi-square = 50.7 p = .001 gamma = .54

Source: Pre-Use Questionnaire
Question: How much time in the average week do you foresee yourself using EIES?

(28 of 89 responding) estimated that they would sign on only once a week or less. Twenty-three of these users in fact became dropouts or low level users (gamma=.50, p=.02).

CONNECTIVITY

There is a weak to moderate relationship for measures of general off-line connectivity to other professionals. For the number of coauthors in the previous year and total number of persons in the specialty with whom the member is in contact, the relationships seem somewhat curvilinear. That is, the isolates and the sociometric stars do not use the system as much as those with moderate numbers of professional connections, who seem to have the most motivation to expand their professional networks. One can categorize the network as a means to upward "social mobility" in the scientific community, for ambitious scientists who were previously at middle levels.

In terms of previous contacts with members of the actual online group, the relationship becomes very strong. The question asked at pre-use was how many persons among those in the specialty with whom the scientist had contacts were in the proposed EIES group. Previously knowing a large number of the online group members is the strongest predictor of very high levels of subsequent use of the system. A series of step-wise multiple regressions was conducted to find the strongest combinations of predictors of amount of use of EIES (see the end of the chapter). When the total number of hours online at follow up was used as the dependent variable (rather than the log of the number of hours, or the level of use, in categories), the strongest predictor is number of participants previously known.

THE EFFECTS OF PERCEIVED COMPETITION

There is no significant relationship between perceived level of overall competition in one's specialty, and amount of EIES use. With the degree of overall competition categorized as intense, moderate, or weak to non-existent, the correlation (gamma) for individuals was only .13 and it was statistically insignificant (p=.72). There is a stronger relationship at the GROUP level (see Chapter 3).

However, there are relationships between perceptions of specific *kinds* of competition and EIES use. Those who perceive competition over funds are slightly more likely to drop out (33% vs. 26%) and less likely to become heavy users in the first three to six months (9% of those reporting competition related to insufficient funds logged 50 or more hours, vs. 18% of those who did not perceive competition of this sort. Overall gamma=.19, p=.13).

Only seven persons who reported competition related to unethical practices among some scientists in the field also completed the follow-up questionnaire. This makes it unlikely that any statistically significant relationships can occur related to reported presence of unethical behavior (which is interpreted as a measure of trust in the group). However, most of those perceiving unethical behavior became dropouts and none became heavy users. The cross-tabulation shows a high level of association (gamma=.68; p=.16). We will call this relationship "strong," even though it does not meet the statistical significance guidelines.

On the other hand, there is a tendency for those who feel that competition in their specialty consists of arguments among those with strongly opposing views to spend more time on EIES. Only 19% of those reporting this reason for competition dropped out, vs. 35% of those who did not. At the other end of the scale, 24% of those reporting opposing viewpoints became heavy users, vs. 10% of those who did not. However, once again we are working with small numbers (21 reporting this form of competition), and even though there is a moderate correlation (gamma=.36), it is not statistically significant (p=.40).

There were no significant relationships with any of the other reasons for competition included in the checklist.

COMPLIANCE PRESSURE

At pre-use, users were asked to indicate whether they were "required" to use the system (only three checked this response), had been requested to do so (a form of pressure), or were free to use it as little or as much as they chose. Pressure to use a system like this seems to have the reverse effect. Among those who felt that they had been requested to use the system, there were more dropouts than among those who perceived free choice, and there were no heavy users in this "non-free choice" group.

FAILED PREDICTORS OF USE

A number of other variables were hypothesized to affect amount of use of a computerized conferencing system. The hypotheses were tested by including indicators of these variables in the pre-use questionnaire, and cross-tabulating them by number of hours online at the time of the follow up questionnaire. Table 2.8 summarizes the statistical results. The following variables are not significantly related to time online (as measured by division into the categories, less than five hours, 5–19 hours, 20–49, 50–99, and 100+):

1. Hours per week spent working on the specialty, or any other reported use of time gathered in the pre-use questionnaire.

TABLE 2 · 8 ***Weak or Insignificant Correlations With Hours of Use***

Question	Gamma	Eta	P
Hours/week in specialty	.03		.47
Number of co-authors in last year		.27	.20
Extent to which scientist considers self in "mainstream"	.12		.11
Total # of contacts in specialty		.31	.23
Frequency of anticipation	−.13		.80
Concern about anticipation	.06		.34
Extent to which emotional commitment governs own behavior	.19		.53
Extent to which emotional commitment ought to govern behavior	.24		.14
Extent to which irrelevancy of personal attributes governs own behavior		.17	.04
Extent to which irrelevancy of personal attributes ought to govern behavior		.20	.02
Sex			.33
Education	.09		.98
Years since highest degree		.22	.29
Books currently in progress		.26	.10
Total articles published during career		.19	.71
Papers presented last year	.16		.06
Total productivity scale	.19		.52
Preference for working in established areas	.17		.49

TABLE 2 · 8 *(cont.)*

Question	Gamma	Eta	P
How well known in field		.21	.62
Whether EIES will affect familiarity with one's work	.31		.68
Reading speed	.12		.86
Typing speed	.17		.31
Preference for writing vs. speaking			.71
Computers are (wonderful/terrible)	.07		.91
Previous use of message system	.11		.13
Previous use of terminals to play games	.14		.45
Access to terminal			.39
Trust computers	.06		.93
Perceived pressure to use the system	.28		.14
Anticipated usefulness of group conferences	.16		.08
Anticipated usefulness of text editing	.20		.01

Source: Pre-Use questionnaire (See appendix for wording)
Notes: Gamma notes linear relationship
Eta denotes curvilinear relationship
"p" is significance level, determined by Chi-square test

2. Frequency of previous anticipation or concern about future anticipation of one's work by others who publish similar things first.
3. Age (There are too few users under 30 or over 50 hours to adequately test this relationship.)
4. Productivity in terms of reported books, articles, etc., either in the previous year or in one's total career. Although correlations

with productivity measures are not statistically significant, there are some moderate correlations. They tend to suggest a curvilinear pattern more than a linear one. That is, those with moderate publication levels before use of EIES tend to use the system more. This makes some sense; those already publishing very heavily probably do not need any new information resources or professional contacts.

5. Preference for working in established areas of science.
6. Subjective report of how well known the member is in his/her specialty. This is contrary to the hypothesis that those who are "low" in the status hierarchy will be more strongly motivated to use the system. However, as will be discussed below, a group aggregation of this variable does have some predictive power—A scientific group seems to need a certain number of "stars" to motivate all of its participants.
7. Whether they thought that use of EIES would affect how well known they are in their research specialty.
8. Reading speed.
9. Speaking vs. writing skill. The question here was whether the prospective user thought that he or she was more effective when writing or speaking. Almost exactly the same proportions of the two types (speakers vs. writers, as self-assessed before system use) became dropouts or heavy users.
10. Typing speed.
11. Attitudes towards computers (either on a "wonderful to terrible" scale, or in terms of trusting them to hold the daily working files that one needs).
12. Previous use of computers or terminals. Neither any of the individual items, nor a combined index on total previous use of terminals was a significant predictor.

There is a suggestion that those who placed a high value on the unique features of EIES as compared to a message system (group conferences and text editing features) are likely to use the system more. This is similar to the finding that expectations about the system's overall usefulness helped to predict hours online.

There is weak support for a relationship between basic values and subsequent use. The pre-use questionnaire contained sets of questions on two of the "pattern variables" used by Talcott Parsons and many subsequent sociologists to characterize value patterns. These are "universalism" vs. "particularism" (whether a scientist or his/her work is judged solely on the basis of their work, or solely on the basis of who they are; in terms of personal knowledge of or relationships with the person) and "affectivity-affective neutrality" (whether a scientist is emotionally committed to his/her theories, or totally objective and not emotionally involved with the theories.)

There are weak relationships showing some tendency for those placing their answers at the "emotional commitment" end of the scales to use EIES more; and for those in the "balanced" area of the choice between the relevancy and irrelevancy of personal attributes for judging scientific work to use it more than those at either extreme. These results are suggestive of possible relationships, but not strong or consistent enough to say that we have proven that such a relationship does exist.

COLLECTIVE GROUP STATUS

Although there is no relationship between the self-assessed status of the individual (unknown to top of field), there does seem to be a relationship with the collective status of the group. As shown in Table 2.9, the groups that had the largest proportion of well known members tended on the average to have the heaviest users of the system. What matters to the individual is how many *other group members* available to communicate with have relatively high professional status.

PRE-USE TERMINAL ACCESS

Although terminal access was reported as an important barrier to use by about 20% of subjects, there was no overall statistically significant association between the terminal access situation reported at pre-use and amount of time spent online by the first follow-up. Many participants

TABLE 2 · 9 ***Status of Group Members vs. Use***

Group	% Hi S	rank	%Hi Use	rank
30	42%	1	46%	1
40	24%	2	34%	3
45	23%	3	17%	4
35	22%	4	43%	2
54	14%	5	10%	5

Note: "HiS" stands for the proportion of group members ranking themselves as 6 or 7 on the seven-point professional status scale.
"Hi use" is the proporation of group members using 50 or more hours of line time by the follow up questionnaire.

were given use of a portable EIES terminal; these were all persons who had reported no access to a computer terminal unless one were provided for them. This meant that they had a light-weight, 30 cps printing terminal available both for office and for home use.

What we find are some puzzling negative relationsips with terminal access and characteristics. For example we find the following:

Home access	*% dropouts*
Report terminal at home	40%
Report terminal available to take home	32%
No terminal available for home	28%

There was likewise no relationship with printing speed of the terminal, though one would be expected. Another puzzling relationsip is that the highest proportion of dropouts occurred among those reporting access to both a CRT and a hard copy terminal, rather than only one. This seems an ideal terminal arrangement for use of EIES. One possible explanation is that those in a terminal-rich environment are also in an already computer-resources rich environment, and do not need additional resources such as EIES.

It is certainly not likely that having a terminal at home or two terminals in the office caused less use of the system, but rather that motivational factors are simply much more important for the scientists in this study. For example, one member apologized for not using the system more because he had to drive about an hour each way to use a terminal—and he was logging over ten hours a month! We have a curious disjunction between the lack of relationship between the terminal access situation at pre-use not being related to level of use of the system, and a fairly important role for subjectively reported terminal access barriers at follow-up. What probably happened is that strongly motivated users with poor access expended the time or money to improve their terminal access situation. But good terminal access alone, without motivation, will not lead to use of the system. In other words, there is an interaction between terminal access and motivational factors. Thus, the overall conclusion reached about the importance of terminal access to system use, given the findings on the follow-up questionnaire as well as the above observations, is that the relationship is conditional on motivational factors. If motivation is weak, poor access becomes a barrier that may be decisive in limiting use of the system; if motivation is strong, users will go to considerable trouble to overcome initially poor access. On the other hand, if there is no access at all, even high motivation cannot lead to high system use.

THE EFFECTIVENESS OF A HUMAN TEACHER

It was assumed that those who had some personal training from another person would be more likely to learn the system and become regular users. Personal training should be more enjoyable. It can be tailored to the questions and difficulties of the individual. There is every reason to believe that the personal teacher should be superior to simply receiving a large, standard document in the mail and teaching oneself.

However, the data do not support this. In fact, there is a statistically significant difference in the other direction—those who had only "How to Use EIES" and the use of online user consultants were less likely to become dropouts or low users than those who had some personal instruction, and slightly more likely to become heavy users.

We do not accept this as cause and effect. For one thing, there are no data about the extent and quality of the personal training that was received. Secondly, it may be that those users who were the most confused and negative were the most likely to seek a personal training session, and that without such personal attention from an experienced user, they would have been even more likely not to accept EIES.

Personal training is expensive and time consuming. The evidence from this study does not justify such expenditures. However, a controlled experiment with random assignment of subjects to different kinds of teaching materials (live teacher, written documentation, online lessons of an interactive nature) would be necessary in order to establish the relative effectiveness of these training methods for different types of users. A current research project is exploring this question. The online lesson may well be the most effective method of all, judging from the many spontaneous requests received from users for this sort of aid. In controlled experiments, first-time users were able to learn to enter and receive material from EIES in about 20 minutes, with an interactive lesson online (see Hiltz, Johnson, Aronovitch, and Turoff, 1980).

SUMMARY OF FINDINGS FOR VARIABLES EXAMINED INDIVIDUALLY

There is indeed a complex causal system at work in determining level of EIES use, as indicated in Table 2.10. When simple two-variable relationships are examined, we find that some characteristics of the individual (pre-use expectations about the system, and preexisting social ties, in particular), of the system and equipment, of the group, and of the task (its relative priority) exhibit moderate to strong associations with level of use of the system. It is also noteworthy that many individual

TABLE 2 · 10 ***Summary of Correlations With Level of Use***

Variable	Correlation*
Individual Characteristics	
A. Attitudinal variables	
1. Attitudes toward media	
a) Attitudes towards computers in general	none
b) Expectations about the specific system	
1) Anticipated amount of use	strong
2) Anticipated impacts on productivity	moderate
2. Expectations about how system use will affect relationships with the group	weak
3. Perceived pressure to use the system	weak (neg.)
B. Work Related Skills and Characteristics	
1. Personal communication skills	
a) Reading speed	none
b) Typing speed	none
c) Preference for speaking or writing	none
2. Previous related experience	
a) Experience using computers	none
b) Use of computer terminals	none
c) Use of other computer based communication systems	none
3. Productivity	
a) Hours per week worked	none
b) Number of publications or other output measures	weak
C. Connectivity	
Number of persons in field with whom one is in contact	weak
Number of persons on system with whom one was in previous contact	strong
"No one" to communicate with	moderate
How well known person is in field	weak
Whether a scientist feels "in the mainstream" or not	weak
Number of coauthors (or coworkers)	weak (curv.)

TABLE 2 · 10 *(cont.)*

Variable	Correlation*
Individual Characteristics (cont.)	
D. Demographic characteristics	
Age	none
Sex	none
Educational level	none
E. Environmental variables	
Position in the organization (or status in informal group)	none
Amount of pressure to use the system (from superiors and peers)	weak (neg.)
F. Basic values (e.g., the pattern variables: universalism vs. particularism, affectivity vs. affective neutrality)	weak
System Characteristics	
In-person or formal training, vs. documentation only	none
Quality of the Telenet interface	moderate
Whether or not the system was a source of difficulties	moderate
System availability (downtime during workday or unavailability nights and weekends)	moderate
Trouble with the telephone or high cost of long distance telephone due to absence of Telenet node	
1. Access to terminal (subjective)	moderate
2. Pre-use access to terminals	
a. Own or share at office	none
b. Terminal for use at home	none
c. CRT, print, or both	none
d. Size and weight and printing speed of the terminal(s) available	none

TABLE 2 · 10 ***Summary of Correlations With Level of Use (cont.)***

Variable	Correlation*
Group Characteristics	
STRUCTURE	
Size	none
COHESIVENESS	
Competitiveness	none to weak
Trust or openness among members	strong
STATUS (are most group members prestigious in their fields, or not?)	moderate
Task Characteristics	
Relative priority of task	strong

* *See footnote 2 for an explanation of the combination of correlation coefficient values and level of significance used in characterizing the strength of observed associations.*

characteristics (such as typing ability, previous use of computers or terminals, and expressed preference for speaking vs. writing) which one might expect to strongly predict level of use have weak or no relationships.

The strongest observed correlate of the level of use is the *anticipated* level of use before experiencing the system at all. This variable is a conglomerate of individual attributes and expectations, probably including basic personality characteristics (not measured), relative importance to the person of communicating with others in the EIES group,, and amount of time available for such activities after the more mandatory job-related tasks are completed.

Measures of connectivity (preexisting communication ties with other group participants) also appear important. An item on the pre-use questionnaire (number of group members previously known) yielded the highest Pearson's correlation coefficient with total hours of use at follow up. An item on the follow-up self-reporting checklist ("There is no one on this system with whom I wish to communicate a great deal") yielded the highest correlation coefficient with level of use of any of the self-reported reasons.

Access barriers as a class (including access to a terminal, trouble with Telenet, and system unavailability) also appear to be important factors in determining the amount of use of EIES. However, it must be noted that with the exception of terminal access, the perception of other access barriers is more an effect of moderate to high use than a cause of dropout or low use behavior: the higher the level of use, the more frequently these barriers were indicated to be "very important."

Among the variables which were hypothesized to be positively related to level of use, but which are not significantly related, are receipt of personal training, reading and typing speed, attitudes toward computers, previous experience with computer terminals or message systems, and how well known the person was in the specialty. On the other hand, groups that were composed of a high proportion of high-status members were, on the average, more active than groups which had a small proportion of well known members.

In order to understand the relative strength and interactions of the variables, those which individually exhibited the most predictive power were entered into a stepwise multiple regression. But before looking at the results of this multi-variate analysis, we will pause to compare the EIES findings with those for a study of NLS.

COMPARATIVE RESULTS FOR A STUDY OF NLS

Gwen Edwards reports extensive data on the correlates of amount of use of NLS, a computer-based text processing and communications system. (See Chapter 1 for a description of the system and the study). We will examine the results in some detail because it is the only other publicly available study which examines a wide range of variables in relation to acceptance of a computer-based communication system.

The variables measured by Edwards (1977) are described in Table 2.11, and her findings are summarized in Table 2.12. Edwards' report frequently gives results for parts of the sample, as well as the whole sample. Results are reported for both total or "general" use, and for just communications use. Sometimes results are reported separately for supervisory and non-supervisory personnel, since this was found to be an important variable affecting use and attitudes. In looking at correlates of usage, the dependent variable "GENERAL USAGE" was broken into three ordinally ranked classes: "low" usage of less than one hour a day (28%); "medium" usage of one to three hours a day (31%); and "high" usage of three or more hours per day (41%). Note that the "middle" level usage of NLS for this study would constitute "high" usage on EIES.

Since Edwards' study was a single cross section, it is difficult to identify

TABLE 2 · 11 *Variables Used in Edwards' NLS Study*

Access—user indicates that there was or was not difficulty accessing the system

Accessibility of work—on a five-point Likert scale, the degree to which the accessibility of the user's work to others is perceived to have increased or decreased

Communications usage—frequency of use of the system for communications purposes (exchange of message, documents, linking in real time)

Direct/indirect usage—direct interaction on the terminal vs. using the system via support staff

General Usage—Total hours per week

Group Incentive—use is required, requested, or the user feels free to use the system as he or she chooses

Home Usage—individual does or does not occasionally use a terminal from home

Initial Perception—the user's retrospective reaction to the system when it was first introduced (thought it would be useless, thought it would revolutionize work/communication processes)

Involvement—the user was or was not involved in the decision to subscribe to NLS

Perception—an index constructed from questions on current perception of the usefulness of NLS (same as initial perception scale, above); and five-point attitude scales on compatibility-incompatibility of the system to normal working/writing/thinking organizing style; flexibility vs. inflexibility of the system; realiability-unreliability of the system

Position—support staff, research, management

Privacy—individual doesn't use the system for work of a confidential nature; takes precautions to ensure the confidentiality of work, such as changing password; or does not let the privacy aspect affect use

Productivity—A five-point scale, the degree to which a user believed his or her work efficiency/productivity decreased or increased as a result of using the system

Professional Image—belief that the system increased or decreased professional image

TABLE 2 · 11 ***(cont.)***

Proximity—the distance between the closest available terminal and the user's office, defined as in the office, within 50 feet, or more than 50 feet from the user's place of work

Quality—A five-point scale, the degree to which a user believes the quality of his or her work has increased or decreased as a result of using the system

Sharing—the individual has sole or shared use of the terminal

Supervision—the user does or does not supervise other employees

Teleconference—the user has or has not ever participated in a teleconference

Terminal Type—teletype only, CRT with teletype version; display based version of NLS with special terminal and electronic cursor

Training—formal program, trained by other employee in charge of training; by other users of NLS; or no training program

Typing Skill—the individual does or does not claim to know how to type

---

cause and effect. For example, when she reports that perceptions of increased productivity are associated with more use, we do not know if there was an expectation of increased productivity before use, the growth of this perception as a result of use, or a combination of both.

Edwards reports that general attitudinal and access variables are most highly related to amount of use of NLS. The strongest correlation (gamma = .69) overall was between use of a terminal at home and amount of use. Typing skill was found to be related to use of NLS only among those who had a negative perception of the system (gamma = .68). Among those with medium to highly positive perceptions of the system, there was no relationship between typing skill and amount of use (gamma = .05). Edwards states that "Once the perceptual barrier is crossed, typing skill is irrelevant to usage." She also suggests that "we can recommend that when implementing an Office of the Future system, it will be beneficial to convince potential users that they need not know how to type to make effective use of the system" (p. 43).

The other variables which are most strongly related to total use are those which indicate perceptions of utility of NLS:

TABLE 2 · 12 ***Correlations (Gamma) with General Use and Communications Use of NLS***

Variable	Genusage	Comusage
Home Usage	−.69	−.52
Teleconferences	−.22	−.50
Professional Image	.50	.49
Terminal Type	.41	.48
Privacy	−.23	−.43
Sharing	−.14	−.40
Productivity	.49	.38
Accessibility	.44	.35
Supervision	−.21	−.30
Initial Perception	.35	.24
Perception Index	.38	.24
Training	.31	.23
Terminal Proximity	.05	−.23
Typing	−.38	.22
Involvement	−.37	−.22
Quality	.38	.12
Group Incentive	−.05	.09
Position	−.10	.08
Direct-Indirect Usage	.18	−.01
Access Problems	−.18	−.01

Source: Edwards, 1977, p. 43. The above correlations have been rearranged in order of strength of association with communications usage. Note that the sign of the correlation depends upon the way the data were coded. For example, if "home usage" of a terminal is coded as "1" and no home usage is coded as "2", then the negtive sign indicates that the home users make more communications usage of NLS.

1. "Professional image": There is a gamma of .50 between the perception that use of NLS will improve one's professional image and NLS use. This is a variable which was not found to be a predictor for the scientists on EIES. A possible explanation is that the opinion of one's organizational peers is much more important to one's future career than the opinions of scientific peers on other academic campuses, who, after all, do not sit on one's tenure or promotion decision-making committee.
2. Perceived impact on productivity: gamma = .49. This is measured with an identical question in the EIES study. The correlations are similar in direction but stronger for NLS.
3. It is related positively to the perception that NLS use increases the accessibility and visibility of one's work to others (gamma = .44).
4. There is a moderate relationship with the user's initial perception of the system and subsequent general use (.35). There is also a moderate relationship with training, and sophistication of the terminal.
5. There is an intriguingly strong negative relationship between previous use of a teleconferencing system and the use of NLS for communication. One can speculate that those who had previously used a system specifically designed for group communication were dissatisfied with the NLS capabilities for this function.

Generally, correlations with communications use are similar to but weaker than those with general or total NLS use. However, one interesting exception is sharing a terminal. It does not affect general use, but having a shared terminal does impact on communications use negatively. Another difference is privacy: concern over it influences communications use much more than general use. These are related, since a shared terminal often means lack of privacy. These differences have an important implication for managers. Using a terminal located in a space shared with other people may be fine when working with data. When composing and receiving text communications, however, privacy is essential. Having to communicate using a CRT in a place with other people is analogous to having one's phone line tapped, in terms of subjective feelings of privacy.

Though results for attitudinal variables measured with the same question are similar for the two studies, there are some contradictory findings for other variables. The explanation may be that the specific questions used were quite different; or, the differences may be attributable to use by an office staff to support their work on the job vs. use by academics to support their informal, organizationally external communication. Still a third source of possible differences in findings are differences between the systems. NLS was a fairly complex, command driven system for augmentation of the individual "knowledge worker," which included some communications components. EIES is primarily a communications system with some text editing, and with a simple menu-driven interface for beginners who have no desire to master the full power available on the system.

The correlations for training and terminal sophistication probably can be explained by the greater complexity of the NLS system for beginners. At the time of the study, it was command-driven, and designed to be used on a sophisticated terminal rather than a simple one. It is not likely that a beginner could learn NLS with no training or personal contact whatsoever with an experienced user. On the other hand, EIES was designed for use on a simple terminal, and to be usable by a beginner in a menu-driven mode without formal training or personal instruction. In other words, the differences for these variables may be attributable to system differences.

MULTI-VARIATE ANALYSIS: STEPWISE MULTIPLE REGRESSION

A stepwise multiple regression was chosen as the best technique for examining interactions among the identified causal factors.[3] The independent variables chosen for inclusion were selected from all those available on the basis of preliminary correlation analyses. They are defined at the top of Table 2.13. Two separate analyses are presented: the first for prediction of LEVEL of use (with 50+ hours as the top category); and the second for absolute number of hours of use. The latter analysis will favor variables which help to predict those with very high hours of use. (A third analysis used the log of the number of hours; its results were almost identical to those for level of use.)

We can see from the correlation matrix in Table 2.13 that the best

[3]The advantage of this technique is that it allows us to compare the strength of the pre-use predictors with that of the self-reported reasons, and to examine interactions among factors that may themselves be highly interrelated. The disadvantage is that the number of cases is greatly reduced; only those who answered all questions on the pre-use and follow-up questionnaires are available for inclusion in the analysis. This reduces our data base to only 65 cases.

A forward stepwise inclusion technique was employed. The order of inclusion is determined by the respective contribution of each variable to explained variance in the dependent variable. The first variable identified by the stepwise analysis is the one that singly explains the greatest amount of variance; the variable that explains the greatest amount of remaining variance in conjunction with the first is entered second, and so forth, until no improvements can be made in the prediction.

TABLE 2 · 13 ***Stepwise Multiple Regression: Determinants of Level of Use***

Variables

LEVEL= Number of hours online at follow-up, categorized as <5, 5–19, 20–49, 50+ ($\overline{X}$= 2.4, *SD*= 1.0).

ESTUSE= Estimated number of hours per week that the system will be used, at pre-use; categorized in six levels ($\overline{X}$= 2.26, *SD*= .91).

NO ONE= Level of agreement with statement at follow-up that "There is no one on this sytem with whom I wish to communicate a great deal" ($\overline{X}$= 2.7, *SD*= .6).

TERM= level of agreement at follow-up that inconvenient access to terminal decreases use ($\overline{X}$= 2.3, *SD*= .9).

KNOWN= Pre-use response, "How many members of your EIES group do you know either professionally or personally?" ($\overline{X}$= 14.9, *SD*= 18.4).

OTHACT= Level of agreement at follow-up on importance of "Other professional activities must take higher priority" ($\overline{X}$ = 1.7, *SD*= .7).

******N of Cases= 65******

Correlation Matrix (Pearson's)

	Term	Known	Estuse	No One	Othact
LEVEL	.25	.26	.46	.29	.02
TERM		.26	.21	.09	.05
KNOWN			.26	.10	.10
ESTUSE				.15	.07
NO ONE					.09

Stepwise Multiple Regression

Step	Factor	Mult R	R Square	Beta
1	ESTUSE	.46	.21	.38
2	NO ONE	.51	.26	.21
3	TERM	.53	.28	.12
4	KNOWN	.54	.29	.11
5	OTHACT	.54	.29	−.04

Step 5 $F = 4.8$, $p = < .01$

overall predictor of level of use is the estimate of number of hours of use per week, made before using the system. In turn, the highest correlate with this estimate is the number of other group members who were already known, before signing on. This gives us some insight into one of the probable strong determinants of this initial estimate—the expectation that the system could be used to increase communication with colleagues with whom one had valued ties.

It is also interesting that although one might expect the "NO ONE" to communicate with factor to be highly (negatively) associated with the number KNOWN before use, there is in fact a weak positive relationship. This suggests that there was a divergence between the expectation of who would be available online, and who actually was there to communicate with. This of course fits in well with the observed high "dropout" rate. In other words, the prospective user knew who was expected to be available online and used this to estimate amount of use of the system; however, many of the anticipated communication partners were among the "dropouts," leaving many group members with the feeling that there was "no one" left with whom they wished to communicate.

The results of the stepwise regression analysis[4] show that after initial estimates of use, the variable which accounts for the most variance in level of use is the "NO ONE" to communicate with factor. This second predictor (NO ONE) raises the proportion of variance explained from 21% to 26%. None of the other variables make much of an improvement in our ability to predict. For instance, although perceived problems with terminal access is selected as the best predictor to be added on the third step, it only increases explained variance by 1%.

For analysis of absolute number of hours of use (Table 2.14), rather than level of use, the number of group members known before the beginning of the computerized conferencing activity is the best predictor. It explains 23% of the variance. We can deduce from the difference between this and the previous analysis that those who knew many other group members before using EIES are likely to use a very high number of hours online, communicating with all of these colleagues. Subsequent steps of the analysis are very similar to those for level of use. Estimated hours improves the prediction significantly, followed by small improvements added by the terminal access and "NO ONE" variables. Altogether, the four variables entered into the equation yield a multiple correlation

[4] In analyzing the results of the stepwise regression, we can look at the order of factors, the extent to which the inclusion of each increases the multiple regression coefficient (MULT R) and its square (R SQUARE, which is the proportion of total variance in level of use that has been explained by the variables included in the equation at each step). BETA is the standardized regression coefficient.

Table 2 · 14 ***Stepwise Multiple Regression: Determinants of Number of Hours of Use at Follow-Up***

Correlation Matrix (Pearson's)

	Term	Known	Estuse	No One	Othact
Hours	.31	.48	.46	.22	.08
Term		.26	.21	.09	.05
Known			.26	.10	.10
Estuse				.15	.07
No One					.09

Hours = Number of Hours on Line at Follow up ($\overline{X}$ = 30, SD = 37.8

N of cases = 65

See Preceeding Table for Other Variable Definitions

Stepwise Multiple Regression

Step	Factor	Mult R	R Square	Beta
1	Known	.48	.23	.35
2	Estuse	.59	.35	.32
3	Term	.61	.37	.14
4	No One	.62	.38	.13

Step 4 F= 9.3, p= <.01

coefficient of .62, corresponding to 38% of the observed total variance in hours online. There is still considerable "unexplained" variance in hours of use, not accounted for by the variables included in this study.

CONCLUSION

We have seen that the strongest predictor of level of EIES use is the participant's own estimate of the time that will be spent online, before ever using the system. This result is more of a puzzle to be solved than an answer to the question of determinants of use. One observed correlate is the number of prospective system members already known. But what other factors account for the formation of such pre-use expectations? Did they hear a presentation on the system, participate in a demonstration, read a book or article? Do the findings imply that it is important to systematically orient and inform users about a system before giving them a chance to sign online? Such questions might be answered with controlled experiments. For example, some prospective new users might be given an introductory lecture-discussion session, while others receive only printed documentation and are told to sign online. Still another possibility is the unmeasured factors of basic personality or work style traits, or perhaps a "hunger" for more communications. Users do seem to "know" ahead of time whether or not they will like this form of communication.

In comparing the findings to a similar study of determinants of NLS use, we found that attitudes and perceptions were important predictors for both systems and types of users, and that typing skills are not a prerequisite for high levels of use. Terminal access and special training were more important for NLS. On the other hand, access barriers such as telephone or packet switching network (TELENET) problems and system unavailability nights and weekends during the first year were moderately important barriers to EIES use, but not included in the NLS study.

The importance of pre-use motivational and perceptual factors suggests that there may be some important underlying psychological or motivational traits that may predict acceptance of systems such as EIES. An indirect indicator of this is the finding that scientists at the middle levels of productivity and connectivity within the specialty tend to use the system more. This may reflect achievement orientation or striving on their part to improve their professional standing.

The findings about the relative importance of social factors rather than characteristics of the system per se in determining acceptance of the medium are similar to those reached by Lucas in his study of management information systems in the office. Based on data on 2,226 MIS users in 16 companies, he concluded that

> Concentration on the technical aspects of the system and a tendency to overlook organizational behavior problems and users are the reasons most information systems have failed (Lucas, 1975: 2).

Whatever explains pre-use expectations or "receptivity" to this form of communication, the practical implications are clear. If prospective conferencing participants do not expect to use the system very much, it is probably a waste of resources to try to put them online. Perhaps CC is like sex in this regard: you enjoy it a lot more if you really want it before you get it, rather than having it thrust upon you.

3.

THE GROUP CONTEXT

The specific group to which an EIES user belongs is consistently associated with reactions to the system. In this chapter, selected findings will be presented which illustrate that perceptions of a computer-mediated communication system and its effectiveness are influenced by the social context of the group process in which an individual participates.

"User group" is a complex variable which includes differences in the following attributes:

1. Nature of the task.
2. Size and social organization of the online research community. This can influence the amount of information flow.
3. Leadership effort and style (or in some cases, lack of any leadership for at least some periods).
4. Special software features which were built for some groups but not for others.
5. Differences in initial attitudes and prevailing norms among group members.

We will attempt to see if any group characteristics are strongly associated with the success of a group in using EIES. Our procedure will be to roughly rank the groups from more to less successful. Then we will look at some characteristics of the various scientific user groups, and see to what extent variations in these characteristics are associated with differences in the level of success. A section which has the theme "The System Is as the User Group Does" follows. It shows how the same system is characterized very differently according to the group membership of the rater. We will also note that considerable "electronic migration" occurred among groups by the end of the operational trials, so that group differences began to blur. Finally, we will consider the implications of the observed group differences for social engineering to facilitate successful applications of such systems for professionals and managers.

A NOTE ON THE COMPOSITION OF EIES GROUPS

The EIES user groups are not necessarily "groups" or "communities" in the sociological sense of having dense sociometric ties, nor did the operational trials groups correspond to the core members of a scientific specialty. One could have created such online "invisible colleges" by starting with citation analysis, locating the leading authors in a specialty and asking them to nominate other members on the basis of their desire to communicate and work together. Instead, a single scientist applied to the National Science Foundation, and nominated proposed participants whom he or she knew to be working in the area. The group leaders (principal investigators) were not necessarily among the best known or best liked scientists within the specialty. As will be seen in detail in Table 3.2, even according to probably over-generous self-ratings of relative status within the specialty, only a quarter of the participants felt that they were at or near the top of their specialties. In terms of people they ranked as major or outstanding in the specialty, most were not on EIES (see Appendix, Preuse questionnaire). The scientists themselves describe their "groups" as more of "a collection of individuals" or "a set of cliques," rather than "a well integrated research community," even at the time of follow-up.

So, in sum, we have collections of scientists working in the same specialty area, most of whom did not know one another before EIES use, rather than true "groups" of scientists.

VARIATIONS IN OVERALL DEGREE OF SUCCESS

There are many ways of measuring the success of a computerized conferencing effort for the members of a user group. We might have gathered a behavioral measure consisting of the extent to which the members felt so strongly about the value of the system that they actively proselytized to bring new members onto the system. We might have counted the total volume of material they created and read. Any one or two indicators are not valid in accurately judging the success of a specific group. Rather, we wish to use a measure that will roughly rank order the various groups as more or less successful in their use of EIES.

We will use two measures of "success" of the operational trials activities for group members. One is the proportion of group members who spent enough hours online that one can presume that they were participating in some activity that they felt to be valuable. The second is subjective ratings of the productiveness or value of the system by those who used it fairly actively. Since most of our subjective ratings of characteristics of the EIES system were included on the follow-up questionnaire, we want to use hours online measures from the same point in time.

Looking first at hours online, we have previously noted that at the three-six months follow-up point, the proportion of group members who had spent twenty or more hours online varied according to the rank order shown in Table 3.1.

We will use as an indicator of the subjectively rated value of EIES the mean rating of group members at follow up of how "productive" or "unproductive" the system was. This was rated on a seven-point semantic differential scale (1 = productive; 7 = unproductive; see Table 3.2).

A technical note first—Group 80 (Hepatitis) is shown in parentheses in the tables because although the same question was used, it was not administered on the standard follow-up questionnaire. Whenever such

TABLE 3 · 1 ***Proportion of Members with 20+ Hours Online at Follow-Up, by Group***

Group	Group #	% 20+	Rank
Futures	30	46%	1
Social Networks	35	43%	2
General Systems	40	34%	3
Hepatitis	80	30%	4
Information Sci	50	25%	5
Devices	45	17%	6
Mental Workload	54	10%	7

TABLE 3 · 2 ***Perceived Productivity of the System, by Group***

Group	Group #	Mean	Rank
Devices	45	2.90	1
Futures	30	2.95	2
General Systems	40	3.24	3
Hepatitis	[80]	3.33	4
Social Networks	35	3.60	5
Mental Workload	54	4.12	6
Information Sci	50	4.40	7

$F=2.2$, $p=.06$

data are available for this group, it will be shown in parentheses, implying we can make some inferences about the relative ranking of group 80, but that the data are somewhat different. The group 80 data are not included in statistical tests of group differences. In the above table, the "F" ratio indicates that an analysis of variance was used to test significance, and the differences among the groups are just short of the rigorous .05 level.

Combining the two kinds of information, we can roughly rank order the groups as follows:

Most successful experiences with EIES = groups 30 (Futures-mean rank 1.5) and 40 (General Systems-mean rank = 3)
Middle level = groups 35 (Social Networks), 45 (Devices), and 80 (Hepatitis) (mean ranks 3.5, 3.5, and 4.0)
Least successful = groups 50 (Information Science) and 54 (Mental Workload) (mean ranks 6 and 6.5)

VARIATIONS IN CHARACTERISTICS OF THE SCIENTIFIC COMMUNITIES

The pre-use questionnaire included many items that could be aggregated to characterize the scientific user groups rather than just individual respondents. It was hypothesized that these characteristics might be important correlates of style and amount of use of the system, and of the outcome of the EIES experience for the groups that used it.

First, let us look at two variables that we would expect to be correlated with group success, given our findings about strong predictors of use for individuals. We have seen that both expected use of the system before ever signing on and the number of group members known before use of EIES correlates highly with one component of group success, time online. However, there is not a strong relationship between these important variables, and group. The differences which do exist among the groups in the number of members who knew each other before using EIES are not significant. The groups in which there were the largest proportions of members knowing one another before use of EIES were in fact not among the most successful—35 and 54.

Looking at expected use of the system before signing onto EIES, the only clear difference is between group 54 (Mental Workload) and the others. Almost three quarters of its members anticipated signing on once a week or less, according to their responses on the pre-use questionnaire. They also had by far the lowest response rate on the pre-use questionnaire (only 8 out of 35, or 23%, whereas no other group had a response rate under 50%). Thus, the least successful of the five operational trials groups on which we have complete data started out the most negative.

It is analytically important that the most powerful individual level predictors of successful adaptation to EIES are generally not confounded with group membership. This means that observed differences in success among groups are due to factors operating at the group level, and are not just spurious effects of differences among individual members.

Table 3.3 shows in more detail a point already covered in the previous chapter. The groups did differ in terms of the proportion of relatively well-known scientists, and the larger proportions of well-known members occured in the more successful groups. Group 30 clearly had the largest proportion of relatively well-known members, while the least successful group (54) had the fewest (only 14%). (Group 50 data are missing for this question and all pre-use measures).

Table 3.4 shows the perceived amount and type of competitiveness, by group. There are some clear differences in perceived competitiveness. Though there is not a one-to-one correlation, the two most successful groups had the largest proportions of members perceiving low or non-existent overall competition. In terms of types of competition, there is a suggestion that fear of unethical behavior among one's peers was most prevalent in the least successful group. However, the number of respondents was so small that the differences cannot be considered significant for that question. Taken together, however, low or non-existent

TABLE 3 · 3 ***How Well Known Participants Were in Their Specialty Areas, by Group***

Group #	30	35	40	45	54	All
1 (practically unknown)	14%	9%	23%	15%	14%	16%
2–3	14	13	27	46	14	23
4 (average)	7	22	17	15	29	17
5	21	35	10	0	29	18
6–7 (tops)	42	22	24	23	14	25
Total	100%	100%	100%	100%	100%	100%
N responding	14	11	9	15	7	65

Source: Pre-Use Questionnaire
Question: How well known is your work, within your specialty area?
: 1 : 2 : 3 : 4 : 5 : 6 : 7 :
Practically Unknown *Average* *Ranked at Top of field*

competitive pressures in terms of perception of intense competition and trust that one's colleagues will not compete unethically are supportive of the success of a computerized conferencing group. On the other hand, perceived competition on the basis of opposing views or theoretical paradigms seems to be healthy for computerized conferencing groups; groups 30 and 40 had the highest reports of this form of competition, and the least successful group had no reports of intellectual competition at pre-use.

NORMS AND COUNTER-NORMS IN THE SCIENTIFIC COMMUNITIES

The norms of science are supposed to stress emotional neutrality and the irrelevance of personal attributes in judging scientific work (See Merton, 1973). That such a scientific ethic exists has been challenged by Mitroff (1974a). Working with Mitroff, two sets of questions were designed to test the perceptions of scientists about the fundamental value commitments which characterize their scientific specialties.

Table 3.5 indicates considerable prevalence of the "counter-norms," and also some differences among specialties. Although the number of respondents to the question is small, the futurists are unanimous in their

TABLE 3 · 4 ***Percent Checking Specific Reason for Competition, by Group***

Group	Funds	Rivals	Drive	Unethical	Opposing
30(N=10)	18%	0%	50%	10%	50%
35(N=20)	21%	55%	65%	5%	30%
40(N=25)	32%	20%	44%	4%	36%
45(N=13)	21%	31%	61%	15%	8%
54(N=5)	9%	20%	20%	40%	0%
Chi square	2.5	11.9	4.5	7.2	7.7
p	.64	.01	.33	.12	.10

Source: Pre Use Questionnaires
Number Responding = 73
Question: What are the reasons for this competition? (Check all that apply.)

Scarcity or competition for funds
Rival groups of collaborators
High achievement or success drive of people in field
Some persons act unethically
Strongly opposing views

opinion that emotional commitment to one's own ideas is characteristic of work in this field. The specialty in which there are the fewest members believing intensely in their own ideas rather than maintaining neutrality until hypotheses are proven is the Devices for the Handicapped area. Even here, commitment is judged much more frequent than neutrality. The two groups which seem to have been the most successful also have the clearest majorities characterizing their peers as emotionally committed rather than neutral (affectivity vs. affective neutrality, in Parson's terms.)

The data on the relevance of personal attributes in judging the work of scientists in the field (not shown here in detail; see Appendix) indicate that the EIES scientific communities believe that personal attributes are taken into account in judging scientific work in their field: "particularism" rather than "universalism" reigns. The only exception is Group 35, social networks theory, where opinion is more evenly divided.

TABLE 3 · 5 ***Whether Emotional Neutrality or Emotional Commitment Governs Behavior of Scientists, by Specialty Group***

Group	Emotional Neutrality More	Equal	Emotional Commitment More
30(N=13)	0%	0%	100%
35(N=20)	30%	20%	50%
40(N=31)	23%	13%	64%
45(N=14)	21%	36%	43%
54(N=7)	29%	14%	57%
Total(N=85)	21%	17%	62%
	Chi-square = 13.7	p = .08	

Source: Pre Use Questionnaire
Question: General Principles of Science
Described below are two sets of conflicting general principles which can guide the conduct and evaluation of scientific research. Please read each set of principles with your specialty area in mind.

Principle A. Emotional Neutrality
Scientists must be emotionally neutral and impartial towards their ideas if they are to stand a fair chance of ultimately being proved valid. Conducting an investigation with anything less than an impartial frame of mind poses the danger that the scientist will bias the results and be unable to give up the hypotheses when they are indeed false.

Principle B. Emotional Commitment
Scientists must be emotionally committed to their ideas if they are to stand a fair chance of ultimately being proved valid. Unless a scientist believes intensely in his or her own ideas and does everything legitimately in his power to verify them, there is the danger that he will give up on his ideas too quickly. Initial inconclusive signs of negative evidence do not warrant a reorientation of research efforts. The scientist must believe in himself and his own findings with great conviction.
On the basis of your own experience and observations, to what extent does each of the principles tend to govern the everyday working behavior of most scientists of your specialty? (Please circle one number.)

A Significantly More Than B	A Moderately More Than B	Both Equally	B Moderately More Than A	B Significantly More Than A
1	2	3	4	5

Responses to 1 and 2, 4 and 5 were combined

Most of the EIES scientific communities do perceive themselves as relatively "new" and unformed. For instance, only in group 45 (Devices) does a clear majority feel that there is an "intellectual mainstream." And only groups 40 (General Systems) and 54 (Mental Workload) have a clear majority feeling that the specialty has been recognized for at least a decade. Whether the prevalence of the "counter-norms" held by these scientists can be accounted for by the relative newness and lack of a mainstream intellectual tradition, or whether the counter-norms might be equally prevalent in older, more established specialties, is an interesting question that cannot be answered by this study.

The final table (3.6) in this section shows that with the exception of Group 35, most EIES users did not prefer to work in well established research areas, but wanted the risk and excitement of working in a new area. The more successful groups seem to have more members who spurn traditional or established areas of scientific work. In fact, if we add together the proportions "disagreeing" or "strongly disagreeing" that they like to work in established research areas, we find an almost perfect rank-order correlation: the highest proportion of "deviants" occurs in the most sucessful group (30), and the lowest in the least successful

TABLE 3 · 6 ***Preference for Working in Established Research Areas, by Specialty Group***

Group	Agree	Neither Agree nor Disagree	Disagree	Strongly Disagree
30(n=16)	6%	6%	56%	31%
35(n=23)	50%	23%	27%	20%
40(n=32)	0%	38%	56%	6%
45(N=14)	0%	43%	50%	7%
54(N=7)	0%	71%	29%	0%
Total	2%	34%	53%	11%

Chi-Square = 18.7 p = .10
Source: Pre Use Questionnaire
Question: I prefer to work in well established research areas. (Strongly agree, agree, neither agree nor disagree, disagree, strongly disagree. Note that no EIES members strongly agreed.)

group (54). This result, like several others in this study, suggests an underlying personality dimension, with those who are more "innovative" in general being attracted to an innovative technology.

TOTAL GROUP CONFERENCE ACTIVITY AND CONTRIBUTIONS BY THE GROUP LEADERS

Table 3.7 shows data on the main group conferences for each group. (Conference 46 is used for group 45 because it was the most active of that group's conferences; Group 50's conference was erased after completion and could not be analyzed.) We see that the groups varied widely in terms of the amount of activity in their main group conference. Group 30, Futures research, was by far the largest conference, and attracted many members in addition to its original participants, once it became one of the most active conferences on EIES. The group leader was extremely active, contributing more than 300 comments to the conference over the course of the discussion.

The second most active conference was 40 (General Systems), according to all measures—the total number of members, total comments written, and number of comments contributed by the group leader.

Thus, the most successful groups are the ones which had the most active conferences. Though group members do many things online besides participate in the common group conference, its success seems

TABLE 3 · 7 ***Group Conference Activity***

Conf	Members	Comments	# by Leader	% by Leader
30	61	1278	312	24%
35	22	289	33	11%
40	45	389	73	19%
46	34	237	52	22%
54	23	138	30	22%
80	11	265	33	13%

Source: Monitor Statistics

central to the perceived quality of the whole experience online.

Group 35 (Social Networks) is something of a special case. After approximately a year of general discussion, about half of the group membership was purged, and a second, task-oriented conference was begun. It is the data for the second of group 35's conferences that are still available for analysis. Group 54 (Mental Workload) had the smallest, least active conference, especially considering the size of its membership. It also had the fewest comments contributed by a group leader.

Group 80 (Hepatitis) is also a special case. This was an explicitly egalitarian task-oriented conference, in which each person had a designated part of the division of labor. Thus, the nominal leader of the expert group contributed only 13% of the items (but 21% of the total lines written; he tended to have longer comments than the average member).

Many factors may account for these variations in activity and apparent success of the main group conferences. One is the extent to which the conference focussed on topics that were interesting and important to the group members. A second, related factor is the level of effort and skill of the group leader. In observing the conferences from week to week, it could be seen that if a group leader went on vacation or otherwise disappeared for more than a week at a time, the conference activity tended to become disorganized and then drop off sharply. The group conferences needed a strong, active leader to keep the discussion organized and moving in a way that was satisfying to the participants. Table 3.8 shows an almost perfect rank order correlation between the leader's effort as measured by time online and our measures of the overall success of the group.

Leadership could be split between two persons. For example, group 80 had a "content" leader who was an expert on the subject being discussed and a "process" leader. Leadership could also rotate. For instance, after the official end of the operational trials, when the leader of group 30 became much less active, conference leadership shifted to another member of the group.

THE ROLE OF A CONFERENCE LEADER

Though managers and professionals would not think of calling a large face-to-face meeting without having someone to be "in charge," many expect a computerized conference to organize and facilitate itself. One of the group leaders has provided an account of his recognition of the necessity for one or more persons to take on leadership roles:

> I for one began with the assumption that a computer conference should pretty much take care of itself. If a group of people with a common

TABLE 3 · 8 ***Leadership Effort and Group Success***

Group	Total Hours By Leader	Leader Hours Rank	Group Success Rank
30	765	1	1
40	755	2	2
45	557	3	3.5
35	320	4	3.5
80	293	5	5
54	129	6	6

Source: Monitor statistics for cumulative time online by Group leader, as of April 3, 1980

> set of interests were given access to EIES, I expected that they could conduct their normal professional communication with enhanced speed and effectiveness. Alas, this was not to be . . . the evidence seemed to support the need for strong leadership . . . It appears that an active moderator is necessary to keep the conference going but that as people get used to the system and initiate their own projects, several leaders begin to emerge. (Umpleby, 1980: 56)

One of the "latecomers" to the operational trials established and led a futures-oriented conference which attracted many members of groups 30 and 40 and other participants after the official end of the Operational Trials. He has documented the role played by a conscientious conference leader, and the kinds of activities which account for the large proportion of comments contributed by the leader in most sucessful conferences (Caldwell, 1981, quoted by permission). Some of his generalizable descriptions of the leadership role are excerpted below:

> The role of a moderator is similar to that of a committee chairman in face-to-face meetings while allowing for the additional unique computer conferencing processes and while not having to worry about some of the meeting characteristics which relate to physical presence.
>
> . . . At the minimum the moderator must enter new members into the conference (once that is done conference members may delete themselves). However, the full responsibilities of a moderator should involve more than this but will vary by conference membership, subject matter, conference activity level, and the personal style of the moderator and members . . .

The conference moderator needs to provide some number of comments which are purely administrative (as opposed to serving as a member of the conference if desired). While the moderator should take the responsibility of making decisions about the conference management it seems reasonable to expect some form of member feedback to assist that decision making. In addition, it may prove helpful to insert certain comments which provide data or literature references on topics relevant to the conference discussion (this could be done by any member) . . .

After the conference had been operating for approximately 30 comments, it was concluded (initiated by member feedback) that an index of every 15 items could help tie things together. In addition, a monthly progress report was provided. After about 50 comments were entered, it seemed advisable to begin an overall index where both old and new conference members could find the other indices. In this overall index there were itemized lists of the separate indices of each 15 or 30 comments summaries as well as locations of the monthly progress reports.

There was also a need for providing directions on conference techniques to several of the members because the backgrounds ranged from members who had started with EIES several years ago and others who had just joined. Accordingly, there were special comments written to provide instructions on how to vote, how to delete comments, how to set conference markers, how to use the associated comment number to advantage, and how to write special commands for efficiency of time. A special conference comment was written which incorporated many of these suggestions and was used to introduce new members to the conference (along with the overall index). Thus, a new member was entered and a message was sent indicating the location of the "hints" comment and the overall index . . .

In this conference, the moderator accounted for 42% of all comments. Most of the time the purely administrative comments (indices, progress reports, member feedback, voting) ran at 20–25% of the total comments . . .

The time devoted to conference management by the moderator depends on the amount of administrative experience, familiarity with the EIES system, and the particular administrative mode required of the conference. In this conference, the development and entering of the monthly progress report took approximately 30 minutes and the indexing of each 30 comments took about 40 minutes. General evaluation of activities and reflections on how to modify conference directions took an estimated three hours per month.

Thus we see two kinds of leadership or medrator activities. There is the administrative support role, which includes admitting and oreinting new members. Then there is the "conference management" role, which includes summarizing the groups's progress in order to focus the discussion, and obtaining feedback from members about the group procedures. These two functions seem critical to a successful conference. In addition, the leader may play an active role in the discussion itself.

RATINGS OF THE GROUP LEADER

The only group where a significant portion of the members listed inadequate leadership as a reason which dissuaded them from using EIES was Group 54. Five of sixteen group 54 respondents to this question (31%) checked this factor as "important."

In response to a direct question which asked "How would you rate the performance of your group leader (principal investigator)?"

excellent 1 2 3 4 5 poor

only groups 50 and 54 have a significant proportion rating the leader below "2." Thus, we may conclude that although there are other factors which also contributed to the lack of success of these two groups, relatively inactive leaders do seem to be an important factor.

VARIATIONS IN SUBJECTIVE REACTIONS TO THE SYSTEM, BY SPECIALTY GROUP

In this section, we will look at how the same objective capabilities or qualities of EIES are differentially perceived and rated, depending on the group context in which a person uses the system. First we will examine the reported responsiveness of people to electronic messages on EIES, which is one effectiveness measure for the EIES message component.

MESSAGE EFFECTIVENESS AND SIZE OF THE ACTIVE GROUP

The message system is designed as a replacement for letters and telephone calls. Of course, it is effective only if members sign on to receive their messages regularly, and answer them rather than ignore them.

The electronic message is generally seen as more effective than or equally effective as a telephone call or letter (Table 3.9). The two groups for which perceived effectiveness is not particularly good are the smallest. It has been hypothesized (Hiltz and Turoff, 1978) that there is a critical mass phenomenon. There must be a large enough number of active members of a user group who sign on daily or almost every day to generate the motivations for all members to sign on frequently and to enter communications into the system. Otherwise, the pattern of daily sign-on which is necessary for such a system to be an efficient (speedy) communication mode is negatively rewarded by "no new items waiting" when a member signs in, and members are discouraged.

However, this question on message responsiveness was not asked of group 80, which was also small, but very committed to their task. Even

TABLE 3 · 9 ***Responsiveness to EIES Messages, by Group***

Group	More Responsive	Less	No Diff.	N
30	55%	20	25	20
35	42%	12	46	24
40	36%	20	44	24
45	40%	35	25	20
50	0	80	20	5
54	14%	43	43	7
all	39%	26	33	101

Chi-Square = 21.2 $p = .13$
Source: Follow Up Questionnaire
Question: When you send a message over EIES rather than writing or telephoning, would you say that recipients are generally
–More responsive to an EIES message
–Less responsive
–No difference

though the total membership was small, a large proportion contributed regularly and was highly motivated. Thus, it seemed to be able to sustain "critical mass" without a large size, though just barely. For example, when the statistics for one specific month (April 1979) were examined, it was found that five of the fifteen Hepatitis group members had spent in excess of two hours online, enough to not only keep up but also to contribute. All except one were active that month, so that they at least could receive and respond to messages. By comparison, group 45 was nominally much larger, with 48 members that month, but only 8 of these spent more than two hours online, and 25 did not sign on at all that month. Thus, it is "effective size" in terms of active members that counts, not "nominal size." In fact, having nominal members who are inactive is undoubtedly harmful to the group, since they do not return the messages sent to them.

Palme (1981) also comes to a conclusion that the effective size of a group is important, but that what is "large enough" varies with the type of task. In a draft report (1981a:17), he states

> The real value . . . will of course depend on whether the person can

> through COM communicate with a sufficiently large number of other persons, with which the person has a need to communicate. How large this group is depends on how large the need to communicate is in the group. If this need is large, and if alternative satisfactory communications means are not available, then COM may work even for a person who only communicates in a small group of e.g., 5–10 people. But for persons who use COM mainly for exchange of experience, communication with at least 30 other people seems to be necessary before this person feels that the value of using COM is larger than the trouble.

The final version of the COM report does not predict a precise number such as 30 as the threshold. Palme concludes (1981b:39) only that:

> One aspect which should be studied more is the optimal conference size. Our experience is that a computer conference seems to work best in group sizes between ten and eighty participants in COM.

THE SYSTEM AS GROUP RORSCHACH

There were consistent though not always statistically significant differences in most other perceived characteristics of the EIES system associated with group. For instance, only members of the user groups that had the least successful field trials on the system tended to frequently feel "distracted by the mechanics of the system." This is probably partially a result of low levels of use, so that the "mechanics" are forgotten between sign-ons, and partially a generalized annoyance with the whole system because of the lack of rewarding activity online. Whether EIES is stimulating or boring, frustrating or not frustrating, is also greatly influenced by group membership (Tables 3.10 and 3.11). Groups 30, 40, and 45 are consistently the most positive; groups 50 and 54 the most negative.

BLURRING AMONG GROUPS

As the Operational Trials proceeded, members of the various scientific communities joined other groups and conferences in addition to their own. Table 3.12 shows, for instance, that of the 24 members of Group 30 online at the end of the operational trials, four were also members of Group 35, eight of Group 40, etc. (Total the numbers above and to the right to read the entire table. For example, of the 66 members of Group 40, seven were also in Group 35 and five were also in Group 45). Migration into two of the groups not included in this study (60, JEDEC, and 70, LEGITECH) is shown for completeness. Groups 30 and 40 had attracted the most cross-memberships with other groups, and Group 54, the least. By this point, many nominal members of the various scientific communities were actually spending more of their online time participating in another group's conferencing activity than in their own.

TABLE 3 · 10 ***EIES Is . . . Stimulating–Boring, by Group (Seven Point Scale; 1 = Stimulating, 7 = Boring)***

Group	1–2	3	4	5–6	Total N	Mean
30	65%	26	0	9	23	2.22
35	56%	24	16	4	25	2.56
40	59%	34	0	7	29	2.38
45	62%	14	19	5	21	2.43
50	40%	20	0	40	5	3.40
54	25%	50	0	25	8	3.13
[80]	58%	17	17	8	12	2.67

Analysis of Variance F = 1.40 P = .23
Chi-Square (uncollapsed data) = 45.6 P = .0007
Note: Only one respondent rated the system at point 6 on the scale; none rated it as 7, or completely boring.

TABLE 3 · 11 ***Whether EIES is "Frustrating", by Group (Seven Point Scale; 1 = Not Frustrating, 7 = Frustrating)***

Group	1–2	3	4	5	6–7	N	$\overline{X}$
30	17%	22	13	35	13	23	4.04
35	20%	12	28	20	20	25	4.12
40	28%	31	21	17	3	29	3.34
45	19%	24	24	29	5	21	3.57
50	0	40	20	0	40	5	4.40
54	0	12	25	25	38	8	4.88

Source: Follow Up Questionnaires
Chi-Square (uncollapsed data) = 4.28 P = .06
Analysis of Variance F = 1.88 P = .10

TABLE 3 · 12 ***Number of Common Members Between Groups April 1, 1980***

Group	30	35	40	45	54	60	70	80
30	24	4	8	3	1	1	5	1
35		25	7	2	0	1	4	1
40			66	5	0	1	9	1
45				30	0	0	5	1
54					21	0	0	0
60						50	1	0
70							36	1
80								16

Source: EIES Monitor data.

In terms of private message activity, an analysis of the EIES "who-to-whom" monitor data by Ronald Rice (1982) shows much the same thing. In the beginning, there is considerable "instability" as individuals and groups seek to establish rewarding information-exchanges. Initially, "group differences" do play an important part in explaining the communication patterns. Increasingly, however, "reciprocity" becomes the best predictor of message exchange patterns: pairs form when senders have their messages returned with answers. Such reciprocated exchanges may be within the group, or, (especially if a person's initial within-group communication efforts are not reciprocated), between groups. Particularly for non-task groups, "The progression over time . . . reveals a shift from groups preoccupied with internal relations only, to groups which interact widely with their electronic environment" (Rice, 1982, p. 10).

SUMMARY

1. The scientific user groups on EIES were collections of individuals and cliques in the same research specialities, rather than "groups" in the sociological sense.

2. A relative group success index was generated using a combination of the proportion of members who spent at least 20 hours online, and subjective satisfaction with productivity gains as a result of using the system. Group characteristics were compared to relative success. We find that:
   a. The least successful groups had the smallest number and proportion of users who demonstrated enthusiasm for the system before ever signing on. They never seemed to build a large enough number of active members to sustain a satisfying group interaction process online.
   b. Intellectual competition within a specialty appears to stimulate use of the system; other types of competition may hinder it.
   c. In terms of the value orientation of emotional commitment ("affectivity") vs. neutrality, the more successful groups have the highest proportion of members who tend to believe intensely in their theories, rather than maintaining neutrality until hypotheses are proven. All of the groups tended toward "particularism" rather than "universalism."
   d. Most EIES scientific users prefer to work in new research areas rather than in well established areas; the more sucessful groups tended to have the largest proportions of would-be pioneers in new research areas.
   e. The amount of online activity by the leader is strongly related to the success of the group.

3. There are correlations between the overall "success" of a group's efforts online, and the subjective impressions of the system formed by group members. For example, the least successful groups are most likely to feel "distracted by the mechanics" of using the medium, and to find the system itself to be "boring" and "frustrating."

   This is an example of the way in which social context mediates the impact of a new technology. The very same technology was perceived differently, and had a greater or less probability of being accepted, depending upon the nature of the user group to which a person belonged, including its values and its leadership.

4. Although users typically join the system as members of a specific group, as they gain experience they tend to communicate with members of other groups, too, and to join other conferences. Initially distinct user groups become overlapping networks. Interesting, well led conferences thrive and grow, attracting members from other groups, while other conferences essentially grow

moribund and the group members stop communicating with one another on a group basis.

User groups within conferencing systems might be compared to subcultures within a society. Being a member of one group (subculture) rather than another seems to shape the experiences of the members and the quality of their (electronic) life.

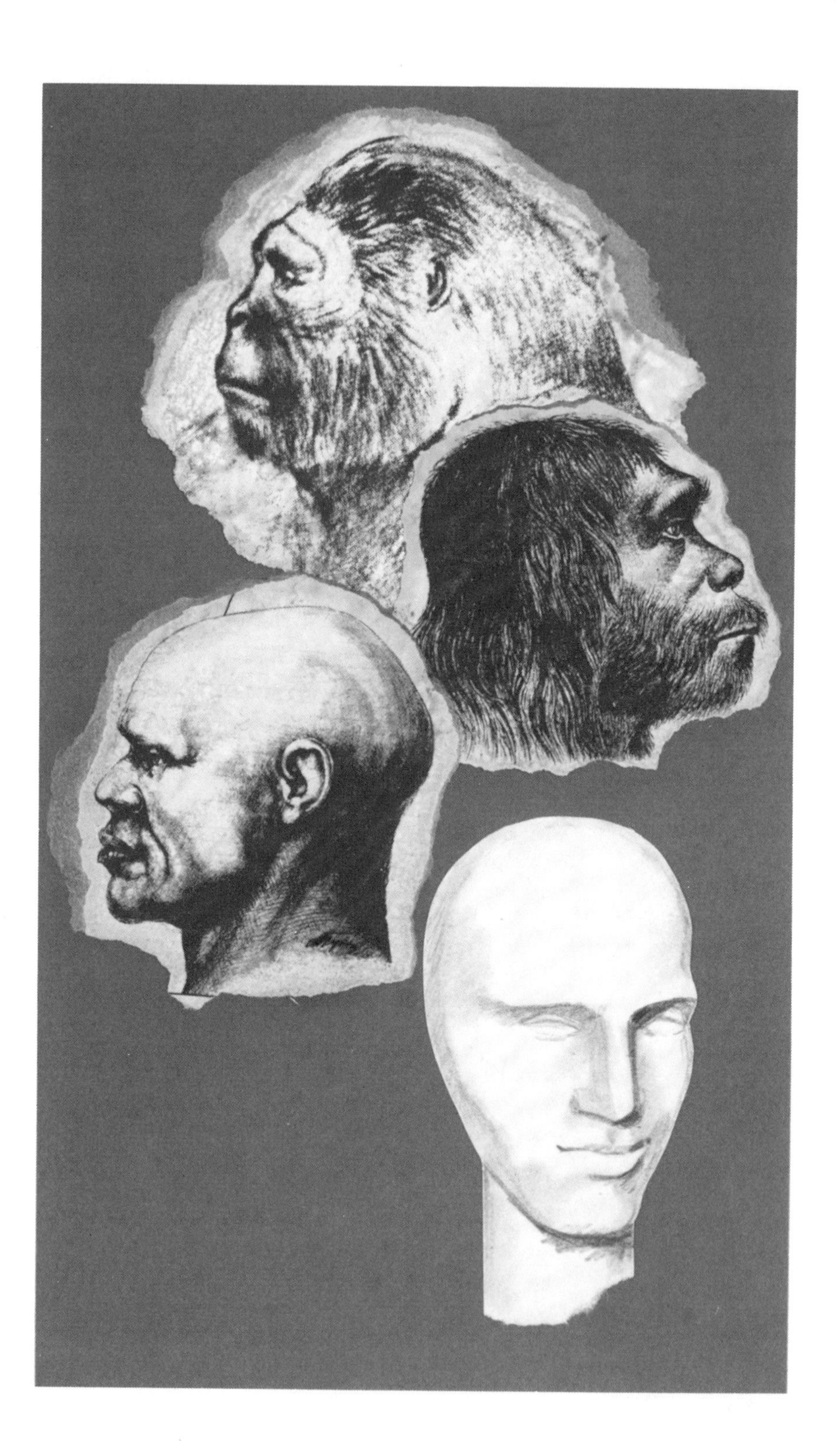

4.

THE EVOLUTION OF USER BEHAVIOR

Coauthored with Murray Turoff

This chapter examines changes in the behavior and attitudes of users in relation to specific features of the system, changes which have some design implications. There are other aspects of behavioral and attitudinal changes of both individual users and user groups over time which are not treated here, such as shifts in perceptions of system usefulness for various communications functions (see Chapter 5), subtle changes in the style and richness of written communications (see Carey, 1980, for a description of paralinguistic behavior), and changes in the social organization and productivity of user groups (see Chapters 6 and 7).

INTRODUCTION

Since the earliest observations, those who have studied computer-mediated communication have recognized that, as Johansen (1976) states, "initial uses of teleconferencing systems often serve as a poor basis for generalizing about future uses." The data from this study provide, for the first time, detailed empirical evidence about changes in user behavior and preferences related to the features or capabilities of computer-mediated communication systems, as a function of experience (hours on-line). They will also serve to show that amount of experience is a powerful determinant of many aspects of user reactions to systems such as EIES.

The basic generalization to be drawn from the data is that there is indeed an "evolution" or pattern of change towards greater complexity, specialization, and diversity of user behavior. This is consistent enough with studies of other teleconferencing systems that it is not likely to be an artifact of the limitations of this study. (See Elton, 1974, and Johansen, Vallee, and Spangler, 1979:136–137 for similar generalizations based upon other teleconferencing systems).

LIMITATIONS OF THE DATA

The reported results are limited to a single system and a single type of user (scientists). Until similar measures are replicated for other systems and other types of users, the generalizability of the EIES results is unknown. Another limitation is that the data currently available for analysis are cross-sectional (attitudes and behavior measured at a single point in time) rather than longitudinal studies which measure each user's amount of experience and opinions about the system at many points in time.[1]

BACKGROUND: THE PROLIFERATION OF SIMPLE ELECTRONIC MAIL SYSTEMS

Computers are increasingly being put to work in the processing, storage, and transmission of text to facilitate human communications. The most widespread proliferation is taking place in the areas of "electronic mail" and "word processing." Uhlig (1977) comes to the same kind of optimistic conclusion about the future importance of electronic mail as do the majority of those who have studied this technology:

> During the next 50 years computer based message systems (CBMS's) will have as great an impact on the way business is done in our society as the impact that the telephone had on business practices during the last 100 years. This, at least, is what our organization has come to believe after two and one-half years of experimenting with them.

Electronic mail is usually designed with a minimal number of features, so that it can simply replicate electronically the delivery of "mail"

[1]We do also have measures of feature ratings at the time of the first follow-up. At that point, there were already significant relationships emerging between time online and features used and valued.

We attempted a longitudinal analysis, but did not have enough cases in the critical ranges for proper analysis. The number of cases for which we have answers on the same question on the value of features on a first follow-up questionnaire at approximately six months after starting to use EIES and on the eighteen-month post-use questionnaire ranges from 55 to 71. However, a total of only twenty were in the range which evolved from fairly new users to experienced users during this time period. Regression analysis and Pearson's correlations on the relationship between change in hours online and change in ratings of features showed relationships that were generally in the predicted direction, but were not statistically significant. We think that the fairly weak relationships are due not only to the low N but also to our inability to capture measures on the users at critical points in their learning behavior. In the future we hope to use automated online questionnaires that will be administered to users when they cross designated thresholds of hours online.

and internal memoranda. For example, this limited set of functions is implicit in the recent paper on the design objectives of message systems by Levin and Schroeder (1979: 29) that refers to "Message systems that communicate memoranda among members of a community." Word processors are likewise being designed as specialized, single purpose systems, to be used only by secretaries acting as intermediaries between the originators and the recipients of text.

In his review of "The Outlook for Computer Mail," Panko (1977) concluded:

> Computer mail has a great deal going for it: apparently favorable economics, a huge potential market, and weakening postal opposition. To tap this market, a fair amount of design evolution will be required.

We agree that "design evolution" will be necessary in order to maximize the role of the computer in the facilitation of human communication. Furthermore, we believe that such evolution should be based upon feedback from the experiences of users of current systems.

LEARNING TO USE THE SYSTEM

THE FIRST SESSION

First time users generally feel a mixture of doubt that anyone can communicate effectively by typing and reading, and lack of self-confidence that they can actually master the mechanics of using the system. They seem to feel a little bit like they are standing alone in the middle of a deep forest and shouting, but nobody is there to hear them. They need to be reassured that indeed their communications are "getting through."

When an account was established, a "greeting" was sent from the user consultants, asking for a return message. This assured that all new users would have a message waiting and a definite objective for their first session, to successfully send a message. It also meant that we could capture some of the feelings of the new users, from the content of their first messages to the user consultants. The most frequent feeling expressed in these first messages is "Somebody talk to me!" They want attention and immediate feedback about whether they are actually communicating, and they feel "strange" communicating in this new way. Examples from first messages in the fall of 1978 are:

"Is anybody out there reading this now? . . . Tell me something. ANYTHING, so I'll know somebody's out there and I'm operating this blank thing right!"

"This is the first message of a wandering wordsmith caught in a time warp."

Until they receive a response to their first message, users do not seem to believe that the system really "works." If they do not manage to sucessfully send and recieve messages during their first 20 to 30 minute session, they may become so frustrated that they hang up and never try again. If they are reassured by receiving some response to their first attempts at communication, they generally proceed on to spend a few hours learning the basic mechanics of the system.

We have identified several stages of learning and adaptation to the system. After mastering the basic mechanics, it takes several more hours to really feel comfortable with this new medium of communication. After that, users slowly begin to explore advanced capabilities, but only as they feel the need for them. In a final stage, which seems to characterize most users with over 100 hours online, they begin to think of new features they would like or new applications of the medium. They have incorporated the system into their daily routine, and generally use the system for what might be termed "eletronic coffee breaks" or "play," as well as for "work."

LEARNING TIME

During the operational trials, EIES could be described as a complex and evolving system which lacked complete, up-to-date documentation and learning guides. As a result, many users felt that it took them too long to learn to use the system, even on a simple level, and that learning the advanced features required too much of a time investment.

At the follow-up, we asked users how long it took them to "learn to use the system reasonably well." This is a rather vague, global question. The reported median is 3.4 hours, but 17% report 10–20 hours, resulting in a mean of 4.96 hours.

At the post-use time, we broke learning down into three levels: the basics, feeling comfortable, and learning advanced features. Most report less than five hours to learn the basic mechanics, though one in five report taking longer. The median is 2.4 hours. There is no relationship between any of the measures of previous experience with using a computer or a computer terminal and the time it takes to learn the basics of EIES.

For "feeling comfortable" the median is 5.1 hours. "Feeling comfortable" is very strongly related to time online when the question is asked (a correlation of .53 as measured by eta, significant at the .01 level). The more time a person had been online, the more time in retrospect they report it took for them to feel comfortable, but the less likely they were to report that they never felt comfortable. Thus, the curvilinear coefficient (eta) is stronger than the linear one (see Table 4.1).

TABLE 4 · 1 ***Reported Number of Hours Online to Reach Various Learning Levels***

Task	<2	2	3–5	6–10	11–97	Have Not
Learn basic mechanics	31%	20	27	14	6	2
Feel comfortable	19%	18	15	14	24	10
Advanced Features	9%	22	10	9	11	38

Source: Post Use Questionnaire
Question: How many hours do you feel it took you to learn the basic mechanics of sending and receiving messages and comments? To feel comfortable communicating with others using this medium? To learn the advanced features that you wanted to use?

Learning of advanced features is a problem area (see Table 5.2). About half of those with less than 50 hours of time online never learned them at all, and one-third of the high users with 50–99 hours online did not. The more time a person spends online, the longer they report taking to learn advanced features. For instance, among those with more than 100 hours online, over one-third report 30 hours or more to learn the advanced features. The difficulty of learning the advanced features and the associated feeling that the system is "too complex" also emerges in many of the optional, open-ended comments about things liked least about EIES and about most needed improvements (see Chapter 5).

An alternative interpretation is that the system is "rich" rather than "too complex." It is obvious that users do not saturate even after 100 hours of experience. The designer points out that a conscious choice is made to let users know that there is an almost endless array of advanced features to be learned if they wish to learn them. By the end of the operational trials, EIES had over 500 commands beyond the menu functions, plus several specialized subsystems and its own programming language.

Included in the Appendix is an example of a report on the questions asked user consultants by EIES members. These show some of the difficulties encountered in learning to use the system. The file of help requests served as one of the main mechanisms for "formative" evaluation, since reported difficulties were used as a basis for modifications to the system or its documentation.

TABLE 4 · 2 ***Reactions to Specific Features of the EIES System and Correlation (Gamma) with Time Online***

Feature	Extremely Valuable	Fairly Useful	Slightly Useful	Useless, Cannot Say	Gamma	P*
Private Messages	68%	22%	10%	1	.50	.09
Text Editing (direct) (e.g., /old/new/)	51%	18	6	25	.23	.47
User Consultants	50%	21	7	22	.32	.02
System Commands (e.g., +cnm)	40%	27	7	26	.49	.01
Group Conferences	39%	33	13	15	.40	.04
Group Messages	35%	31	25	9	.06	.48
The Directory	34%	35	17	14	.21	.04
Private Conferences	33%	25	8	35	.44	.01
Retrieval	31%	31	9	30	.30	.48
Searches	27%	16	18	38	.38	.01
User Defined commands (i.e. +Define)	21%	15	5	59	.29	.001
Text Editing (indirect) (e.g., .text)	20%	16	3	61	.17	.16
+SEN and ???	18%	21	10	51	.58	.001

Chimo	17%	23	24	36	.34	.20
Private Notebooks	14%	23	7	56	.42	.001
Use of ?,??	12%	25	16	47	.11	.24
Explanation File	10%	20	19	51	.00	.82
Terminal Control Features (e.g., +left, +page)	10%	17	7	66	.22	.19
Anonymity or Pen Name	10%	13	16	61	.32	.25
Synchronous Discussions in Conferences	9%	12	16	63	.17	.65
Group Notebooks	7%	15	6	72	.03	.39
Special Programs (e.g., +terms, +respond	9%	9	6	76	.40	.12
Graphics Routines	7%	5	2	86	.42	.21
Interact Programming	5%	3	6	86	.20	.16
Tailored Interfaces (e.g., +Legitech)	4%	6	3	87	.41	.03
Games (e.g. +story)	3%	6	21	70	.55	.002
Voting	2%	12	7	79	.18	.15

Source: Post Use Questionnaires, N = 102
**Probability that relationship could be due to sampling error, Chi Square test*

EVOLUTION OF USER BEHAVIOR

After approximately eighteen months of use of the EIES system, members of the scientific user groups online were asked to rate the perceived usefulness of a number of specific system features. If they had not used a feature at all, they were instructed to check "Cannot Say;" otherwise, they were to rate each one as "Extremely Valuable," "Fairly Useful," "Slightly Useful," or "Useless." (See Chapter 1 for descriptions of these features).

The data in Tables 4.2 and 4.3 show the relationship between amount of time spent online and system feature ratings. In Table 4.2, the first column served to order the features, and is simply the proportion of the 102 users answering these questions who rated a feature as "extremely valuable." The responses at the other end of the scale, "Useless" and "Cannot Say," have been combined to form a more nearly ordinal scale, since very few checked "useless." "Cannot say" was the response that was checked by respondents who felt so little need for the feature that they did not ever try to use it. Some of this is accounted for by poor documentation of the newest of the features, which are not included in the user manual.

Column five of Table 4.2 reports gamma, which shows the strength of the relationship between the subjective rating of a feature, and amount of system use at the time the questionnaire was answered. The last column shows the level of statistical significance of the relationship between time on the system and subjective ratings of the value of the features, based on a Chi-square test.

The most universally appreciated features are the private message, the direct text editing necessary to make typing corrections, user consultants to help one find one's way around the system, and system commands to replace a menu-driven interface when users understand the options available. These are the types of features which are built into most electronic mail systems, with the exception that most such systems do not include the "friendly human helpers," the user consultants. However, high overall popularity ratings are also received by many features which are not usually part of electronic mail systems: group and private conferences, and the public directory of members to facilitate the formation of interest groups. In addition, we notice from gamma statistics that appreciation of many features appears to be related significantly to amount of use of the system.

This becomes clearer in Table 4.3. Here we see that beginning users do indeed see the need for only a relatively small number of features in a computer-based communication system. However, the more experience they gain, the more they come to feel that a wide variety of communication spaces and capabilities is necessary, and the less likely they are to be satisfied with a simple message system. The group-oriented and conferencing features become much more important, as do the features that are necessary for storage, retrieval, and manipulation of text for documents.

EIES is, admittedly, not what it should be in terms of user documentation. As an R&D system with low levels of operational staff, there is no regular documentation effort. New features arise from user feedback via the user consultants and evaluators to the implementors. When a new feature is added, it is exposed to the user consultants, who test it and write documentation for the online file. Major new features are announced in CHIMO, the online newsletter. After that, a user must either search the explanation file or ask a user consultant if a feature exists to fill a perceived need. There is no regular mailing of updated documentation to users. As a result, a user must feel motivated to seek out new features and to learn to use them without any face-to-face training. We think that the users themselves seeking out new features after gaining experience online makes our results more significant than they would be if they were simply responding to pushes from "advanced training seminars" or published training manuals on the features which they "ought" to learn when they feel comfortable with the basic system.

Although the likelihood that a person will find a system feature necessary or useful is generally positively correlated with use, there are a few exceptions. Some of the features for which perceived usefulness seems to be a direct function of amount of use of the system are: group messages, group conferences, private conferences, system commands (as compared to the menu selection interface), search routines, and indirect editing for formatting of output.

One interesting drop is in the perceived value of group messages, at the intermediate levels. We think that new users perceive the feature from the point of view of the sender: a convenient way to communicate with a large group. With a little more experience, however, they become aware of unwanted group messages from the recipient's point of view. Group conferences, in which receipt of an item is governed by self-selection on the basis of topic, are then seen as a more valuable, self-filtered mechanism for group communication, within the context of the EIES design.

An interesting curvilinear pattern occurs for user consultants; appreciation of them is high at all levels, but the newest and the most experienced users find them most valuable of all. This is probably because the user consultant is asked for help and human response ("Somebody talk to me!") by neophytes, and then becomes the source of "advanced knowledge" on features that are too new or complicated to be auto-

TABLE 4 · 3 ***Growth of Features Perceived as "Extremely Valuable" or "Fairly Useful" as a Function of Amount of Experience Using EIES (* indicates addition to list over prior usage class)***

Users with 1 to 19 Hours Online (N=26)

Feature	%
Private Messages	81
User Consultants	71
Group Messages	68
Group Conferences	58
Direct Edits	63
Membership Directory	59

Users with 20 to 49 Hours Experience (N=32)

Feature	%	% Shift
Private Messages	84	+3
Group Conferences	66	+8
Direct Edits	65	+2
System Commands	64	+21
User Consultants	59	−11
Group Messages	62	−6
Retrieval*	53	+5

TABLE 4 · 3 *(cont.)*

Users with 20 to 49 Hours Experience (N = 32)

Features	%	% Shift
Private Conferences*	53	+17
Membership Directory	56	−3

Users with 50 to 99 Hours Experience (N = 25) *(cont.)*

Features	%	% Shift
Private Messages	96	+6
Group Conferences	80	+14
System Commands	75	+11
Membership Directory	72	+16
Retrieval	68	+15
User Consultants	67	+7
Direct Edits	67	+1
Group Messages	54	−8
Searches*	52	+26
? and ??*	52	+10
Private Conferences	51	−2
Send, Link, and ???*	50	+26

TABLE 4 · 3 ***Growth of Features Perceived as "Extremely Valuable" or "Fairly Useful" as a Function of Amount of Experience Using EIES (* indicates addition to list over prior usage class) (cont.)***

Users with 100 Hours and Over Experience (N = 19)

Feature	%	% Shift
Private Messages	100	+4
Membership Directory	95	+23
User Consultants	95	+28
Direct Edits	90	+13
Group Conferences	90	+10
System Commands	90	+15
Retrieval	84	+16
Group Messages	84	+30
Private Notebooks*	74	+44
Sen, Link, and ???	79	+29
User Defined Commands*	68	+31
Chimo*	63	+42
Indirect Edits*	63	+34
Private Conferences	55	+4
Terminal Control*	53	+46

Source: Post Use questionnaire and Monitor Data on Accumulated Hours.

matically retrievable by the short explanation request (? and ??). This tends to occur when users master the basic system and are ready to move on to preparing large documents in notebooks and defining their own commands.

Another complementary explanation, partially verified by observation, is that the user consultants also take on gatekeeping and information brokerage roles. They are often asked by advanced users for information on where particular topics might be discussed and who else on the system might be interested in them. In a sense, the user consultants represent a new type of human facilitation role for the electronic information exchange environment. They also advise on effective styles of leadership for users who wish to establish a conference or other activity online.

Looking at the pattern of changes, one can interpret them as showing that new users appreciate a system that replaces communication media with which they are familiar. These are the letter and the telephone call (replaced by the private message), and the meeting (replaced by the group conference). However, as they gain experience with the new medium, their perceptions of useful applications and their preferred styles of using the medium change.

As users gain more experience with the medium, they tend to find more valuable the unique kinds of functions which the computer can provide for asynchronous group efforts. They need features which help them to deal with "information overload," which can result from intensive daily interaction with a large number of people and groups. They also begin to use other advanced features that can be provided by a computerized conferencing system.

Features for which there is a substantial increase in perceived usefulness as a function of experience can be classified as:

1. Features that facilitate long-term group communication rather than one-to-one communication (the group conference and the private conference).
2. Features that allow a user to actively control the system rather than passively react to menu choices and new items automatically presented (system commands, user-defined commands, searches). However, it should be noted that EIES members feel that the menu is the optimal interface for the beginning user.
3. Features to support composition and the preparation of larger text items and documents (notebooks, indirect editing, and terminal controls for formatting output). Note that it is only at 100 hours or more of experience that most users arrive at the point where they want to produce their large documents online, rather than having them typed.

4. Features that permit tailoring of the system to individual and group needs (user-defined commands, special routines, and the INTERACT language).

PHASES OF USER BEHAVIOR

One classical model of user behavior in interactive systems with which one can compare our data was developed by Bennett (1972). He generalizes user behavior into the *uncertainty* phase, during which the learner has to overcome hesitancy and anxiety; the *insight* phase, during which the user understands the general concept of the system and can make at least limited use of it for his or her own purposes; the *incorporation* phase, when the mechanics of the interaction become second nature; and the *saturation* phase where the system is perceived as inadequate for meeting new requirements users evolve as a result of experience.

EIES users report a median of 1.7 hours to learn the basics, but there is quite a wide variation (the mean is 6.4 hours). Reaching the "Insight" phase seems to be related to becoming comfortable with the writing style and multi-strandedness of conferences, where many topics tend to be discussed simultaneously. A median of 4.5 hours is reported to feel comfortable using the system. The "Incorporation" phase appears to have occurred by 50 hours. To date, we have not observed any signs of the "Saturation" phase, except in the form of a desire to learn the INTERACT programming language and construct one's own subsystems, or have another person do the programming to specifications of the users.

There is a phenomenon of "information overload," which seems to set in on all regular users sooner or later. EIES provides many conferences and activities which users are free to join, far more than the number with which any individual can cope. The growth in publicly available conferences and the fact that a new user can go back and read a conference transcript that has been accumulating for a year or more makes the accumulated material in EIES like a data base. The plethora of available material creates a need for searches, retrieval, and the ability to select material of interest from all that is stored online. This overload phase is now receiving considerable attention in the evolution of the EIES system design.

PLAYING ONLINE

Once users have gotten comfortable with the system, they tend to use it for socializing as well as for task-oriented activities. They exchange gossip and pleasantries, support and comfort one another at times of personal crisis, look for interesting activities online, flirt, and invent new forms and applications. Births, deaths, and marriages have been announced, and friendships formed. For example, about a dozen users during the operational trials invented a "new art form," an "electronic soap opera." Participants "became" a character in the story, taking on a pen name for this purpose. Each character pushed the story forward with his or her entry. It was a science fiction soap opera, complete with clones, space-time travel, and technological marvels. Other examples of institutionalized "electronic coffee breaks" are the public "Poetry Corner" and "Grafitti" conferences, where amateur poets publish poems and (some would say, definitely amateur) comics and pundits tell jokes and make wry observations. Much of this "play" activity is done using pen names or anonymity. It appears to make important contributions to improving the social cohesion of the online communities and to making the system enjoyable to use.

INCORPORATION OR ADDICTION?

Some of the users who have spent a lot of time online and have incorporated EIES into their style of work refer to themselves as "addicted," or make comments that could be interpreted as signs of dependence. Some examples of this are:

> "I can't think when the system is down."
>
> "I can live without EIES, but I can't LIVE without EIES!" (conference 1003, Impacts)
>
> (In explaining reasons limiting use on the post-use questionnaire . . .) "The only pressures were the need to sleep and to continue the obligations of a life that already consumed 16 to 18 hours a day. But for that I would have signed on EIES regularly once a day, for 16 hours each time."
>
> (In response to post-use questionnaire item on changes in the way one works and thinks . . .) "I spend 1–3 hours per day on EIES, usually in the morning, often on weekends and at night. It has become 'essential' to me."
>
> "During and after the Berlin WFS meeting I became somewhat addicted to EIES."
>
> "I find myself staying up late at night and getting up early in the morning just to use the damn thing."

A study of "ex-addicts" had been planned as part of the extension for this study. The operational trials groups were ended on April 1, 1980. The plan was to study users who had spent more than 100 hours online, at approximately one month and six months after their last use of EIES. However, practically all of them managed to find the funds somewhere to continue EIES membership, if only as "class 2" members paying out of their own pockets. So we have no "cold turkey" behavior to report.

DESIGN IMPLICATIONS

Given the observed evolutionary patterns, short-term pretests of inexperienced users on small-scale systems cannot be generalized to predict the preferences of experienced users on operational systems. Users cannot tell you what they need prior to using this technology. Attempts to pre-design fixed systems, which are common in the standard data base area, are doomed to failure, unless the group setting the requirements are experienced users of the technology. The difficulty in validating this statement is that people in dire need of improved communications will utilize anything they are given which provides increased efficiency. Simple message systems will do this, but they will also leave the user in ignorance of other opportunities which this technology can offer.

Our experiences with EIES suggest that system functions should be modifiable and extensible for specific groups and applications. Depending upon the type of application and experience level of a user group, different styles of interface and function-specific features designed to support specific types of tasks can greatly aid satisfaction and acceptance. By *modifiable* systems, we mean software which can easily be changed based on user feedback, and which can be made to appear or operate differently for specific individuals and groups within the user population. By *extensible* software, we mean systems that include a programming capability to add specific kinds of communications capabilities, decision support structures, or other modules which will help to support specific kinds of applications. The argument for evolutionary design is treated in detail elsewhere (Hiltz and Turoff, 1982). However, we do wish to include here an overview of the contrast between a group that participated in an "evolutionary design" approach, and one that did not.

THE ELECTRONIC JOURNAL: A FAILED IMPLEMENTATION

It is said that one can learn from failures as well as from successes. The disgruntled Mental Workload group (see Chapter 3) is perhaps a case in point. The task which this group undertook was to create and implement an online replication of a scientific journal. Specifications were set before the group ever tried using the system. The electronic journal was to replicate, in so far as possible, the processes typical for a printed scientific journal. Software was created to meet the specifications arrived at before the group had any extensive experience with the medium. The software worked, but it was a failure, in the sense that practically nobody used it. There was no provision made to modify the software as a result of experience; to bury the stillborn system and start over again. This group had many other problems (see Guillaume, 1980), but failing to provide for an evolutionary and participatory design process to meet their needs was not the least of them. In addition, they were missing the second component of what has been termed "groupware": this component is facilitation of a group process which would encourage the active participation of members in their online tasks.

> The types of activity observed and the continuing lack of social and procedural interactions suggest that the failure to produce a journal was not a result of the hardware and software aspects of the system, but rather a result of the failure of the group to recognize and apply appropriate maintenance and task functions which would have facilitated the work of the group . . . The failure, then, was a result of a breakdown in group processes. (Guillaume, 1980:27).

PARTICIPATORY EVOLUTIONARY CUSTOMIZATION: THE SUCCESSFUL CASE OF TOPICS

The combination of terms in the subtitle, though somewhat awkward, seem to all be necessary to describe what Tapscott et al. refer to as "user-driven customization of integrated office systems to a specific user environment" (1980: 1). We originally referred to it as "structuring the group process" (Hiltz and Turoff, 1978). Peter and Trudy Johnson-Lenz have invented the term ""groupware" (Johnson-Lenz, 1981b) and given a complete case history of an example of the evolution of successful groupware (Johnson-Lenz, 1981a); we will rely on their descriptions to illustrate what we mean.

> A group working together in a computerized conferencing environment, following certain procedures, can be greatly aided by software which supports and facilitates those procedures. However, software procedures are only one component of structured communication. The other major component is the processes and procedures used by the group. The most effective use of the medium comes about when a group uses processes and procedures specifically designed to meet its needs, plus computer software which supports and facilitates those procedures . . . This union of GROUP process and computer softWARE support we call GROUPWARE . . . (Johnson-Lenz, 1981b)

An unstructured message exchange or computerized conference can work well if a group is not too large and not discussing too many topics or tasks at once. For large groups and/or multi-faceted tasks, however, something must be done to order and filter the exchange of information if the members are not to suffer from immense information overload and confusion.

The TOPICS groupware was orginally designed as a special-purpose structure called LEGITECH in 1978, to support a many-to-many inquiry-response network among state legislative science advisors. A print and telephone-based system had already been in place and the initial design attempted to automate this process. The terms used are INQUIRY, BACKGROUND, and RESPONSES. An INQUIRY is a short "all points bulletin" which anyone in the network can broadcast to anyone else. The BACKGROUND statement is a much longer description of the question and related issues; the original inquirer composes this longer statement, and members of the network may choose whether or not to read it, based on their own interests and available information. Anyone with information to share may RESPOND to the inquiry; other members of the network interested in the same inquiry topic may elect to read these responses. (Stevens, 1980).

The initial design was created by a small design team of which the group leader or manager (Stevens) was an active part, and aimed at translating an off-line process into an online communication structure. So far, the process is much the same as that followed for the electronic journal software. However, TOPICS went through three subsequent versions over more than two years; and each stage in its evolution was based on a great deal of feedback from its users about problems and desired enhancements; thus its development represents not only a "customized" structure, but one that was designed in an *evolutionary* and *participatory* manner.

The resultant TOPICS system was so robust that it not only survived, but spawned a whole family of topics exchanges which served several overlapping networks of users. A version of TOPICS was eventually transferred to THE SOURCE for widespread commercial application.

DESIGN CONFERENCES

In addition to the TOPICS system described above, several other user groups have conducted design conferences on EIES to create the initial specifications for a tailored subsystem and its subsequent modifications. In such discussions, there is a specific goal, division of labor, and timetable, however vague. The stored transcript facilitates a kind of "group memory." Users and systems designers generate and prioritize their needs and arrive at initial system specifications. After the initial version is implemented, discussions focus on its adequacy for meeting original and changing needs, and enhancements or modifications are arrived at.

DESIGN CONCLUSIONS

The need for user involvement in system design has been recognized for MIS systems:

> If the user initiates a system he will have more commitment to design, implementation, and use . . . Let the user design the system if possible. The information services staff should act as a catalyst and map the user's functional and logical design into manual procedures and computer programs" (Lucas, 1975: 111–112)

This is no less true for computer-mediated communication systems. The users are the "experts" on their communication needs. However, only with experience do they become informed about what kinds of structures or features are possible with this new form of communication, and more importantly, what kinds of features will be really useful to them in managing their interactions online.

SUMMARY

Our data do not enable us to completely disentangle the causal sequence of acquired experience with CC and appreciation of a wider range of system features. Ideally, future studies will capture behavioral as well as attitudinal data on changes in feature use as a function of time online. Our findings do indicate that experience, as measured by hours online, is likely to be related to many patterns of system use and attitudes toward it. Therefore, experience will be used as an independent or control variable in looking at dependent variables in subsequent chapters.

When a user first signs on, he or she is likely to feel that this mode of communication is "unnatural" and difficult, and to feel awkward and inept. Sucessfully "getting through" spurs the motivation to learn the basic mechanics. After several hours of experience, the mechanical procedures become transparent and one begins to feel comfortable with the system. As the volume and complexity of their interactions online increase, users are motivated to learn advanced features which enable them to generate and organize their communications more efficiently. They also begin to actively explore the potentials of the medium and to ask for features that do not seem to exist. Many of the experienced users, with 100 or more hours online, begin to refer to themselves as "addicts," in the sense that they not only have incorporated the system into their daily pattern of activities, but also get a "kick" out of using it.

Because user behavior changes over time, computer-mediated communication systems should be designed to facilitate user participation in the evolution of system features to serve emerging needs. We can also recognize a conflicting set of design principles: new users want a system that seems simple, but experienced users want a rich, responsive system that enables them to take an active role in tailoring it to their needs and preferences.

Perhaps CC systems are more like wine than roses. "A rose is a rose is a rose . . ." But experienced users of CC, like oenophiles, come to appreciate design subtleties and complexities, and to want to be able to choose just the right feature to support or complement a variety of communication activities.

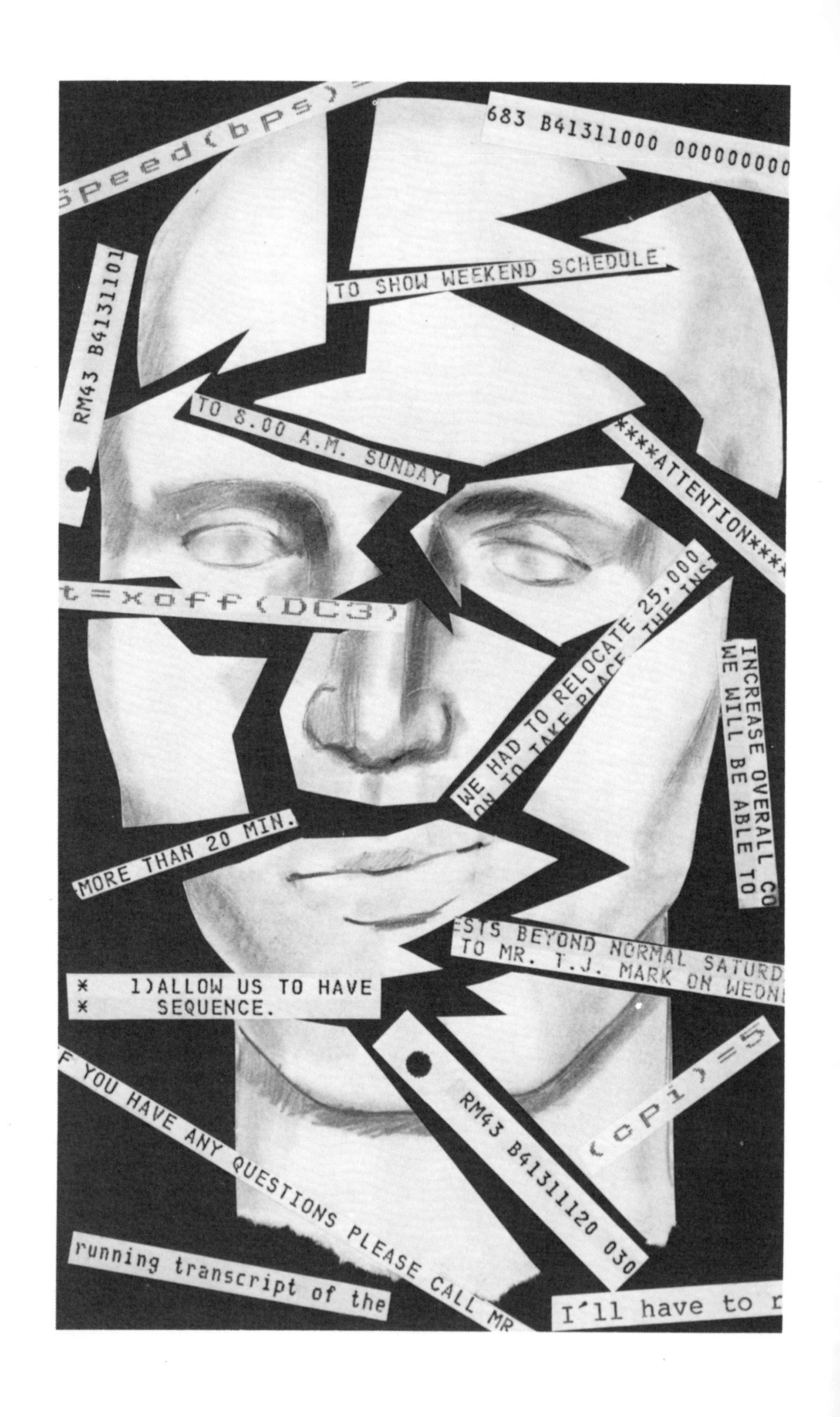
Speed(bps)=
683 B41311000 000000000
RM43 B41311101
TO SHOW WEEKEND SCHEDULE
TO 8.00 A.M. SUNDAY
ATTENTION
t=xoff(DC3)
WE HAD TO RELOCATE 25,000
ON TO TAKE PLACE THE INS
INCREASE OVERALL CO
WE WILL BE ABLE TO
MORE THAN 20 MIN.
ESTS BEYOND NORMAL SATURD
TO MR. T.J. MARK ON WEDNE
* 1)ALLOW US TO HAVE
* SEQUENCE.
F YOU HAVE ANY QUESTIONS PLEASE CALL MR
RM43 B41311120 030
(cpi)=5
running transcript of the
I'll have to r

5.

REACTIONS TO THE SYSTEM

This chapter examines users' opinions and experiences in considerable detail. We begin with use of the documentation and help features, and how users deal with the output they receive. Then we proceed to reactions to specific features, particularly the interface, and those aspects of the system that are most and least liked. Overall system evaluations are looked at both in terms of global characterizations such as being "good" or "bad," "stimulating" or "boring," and in terms of perceived utility for specific communication and information exchange functions. The reactions of other family members who are indirectly affected by the system will also be described. Finally, a multivariate analysis will be used to construct satisfaction indices and to identify the most important determinants of subjective satisfaction.

DOCUMENTATION AND HELP FEATURES

New users of EIES were provided with a loose-leaf red binder called "How to Use EIES." It covers the basic features and includes a one page "users' guide" that is a map of the system, a list of frequently used editing symbols, and a "quick start guide."

Ratings of this documentation by users are positive. Almost all find it readable and fairly easy to understand. However, this written documentation does not cover the new and changing features of EIES, or the advanced features. Moreover, as with all printed documentation, there are a significant number of people, especially among those experienced with interactive systems, who choose not to read it. In some cases, there seems to be the attitude that one can instinctively master this system in an exploratory manner, and that having to read a formal manual would be insulting and dull. Thus, it was necessary to develop sources of online help to supplement and update the written documentation.

An online explanation file with extensive search capabilities was constructed to serve as a comprehensive, constantly updated source of information. The follow-up questionnaire from spring 1978 showed that very little use was being made of the explanation file. Only 2% rated it "extremely valuable," and 30% said they could not say anything about it because they had never used it.

We then modularized the online instruction by introducing two new features. ?WORD (i.e., ?message; ?edit) gives a paragraph to one page explanation of any feature on EIES and can be entered at any point. Second, a system of short explanations was linked to all of the choice points. One question mark (?) in response to an EIES query produces a one line explanation of possible responses. Two question marks evokes a paragraph or two of instructions. Thus, whenever a user does not know what options are available, documentation can be easily retrieved. Much of the material retrieved by ?word and ?? comes from pieces of the explanation file.

An online newsletter was also developed to keep users up to date with short stories about new features, groups, and activities. A naming contest resulted in "CHIMO," which supposedly means "I am your friend" in Eskimo language and is a sort of networker's expression of solidarity.

"Last but not least," there are the human "user consultants" available on EIES. The user consultants are volunteers who receive accounts and TELENET time in exchange for playing multiple roles as go-betweens for the system and its users. They supplement the printed and online documentation in helping both new and advanced users to learn how to use various parts of the system. They provide a human source of support and encouragement and serve as peoplebrokers in assisting users to find others interested in the same topics. In addition, the user consultants test new features and actually write the documentation for them; these functions are generally not visible to other EIES members. In sum, the user consultants play roles analogous to marriage counselors, brokers, and therapists.

As a result of an evolutionary design process, EIES offers a wide variety of means of helping users who need to learn about the system, in addition to the written documentation. Since users are free to choose any combination of the aids, their relative popularity may be of interest to other designers of interactive computer systems in deciding which types are most important to include.

TABLE 5 · 1 ***Percent Making Frequent or Occasional Use of Online Help Aids, by Time Online***

Feature	Hours Online				All	Gamma*	P
	<20	20–49	50–99	100+			
User Consultants	80	67	83	95	79	.26	.29
CHIMO (News)	56	67	87	89	73	.47	.02
? or ??	43	50	62	56	52	.11	.55
Explanation File	70	43	48	65	47	.18	.28
N	25	33	24	19	101		

Source: Post Use Questionnaire
**Gamma is a measure of association between time online and frequency of use.*
P is the probability that the results could be attributed to sampling error, based on a chi square test.
Question: Checklist with categories "frequently", "occasionally" or "never" for frequency of use.

USE OF ALTERNATIVE HELP FEATURES

Table 5.1 shows the reported relative frequency of use of the various online help aids. The most popular are the human "user consultants," described in detail below. Next most widespread use is made of the online newsletter ("CHIMO"). This is followed by the "?word" system, and the full explanation file is least used. However, all of them are used frequently enough so that ideally, a system should incorporate the full range.

THE USER CONSULTANTS

The user consultants are very popular. In the post-use checklist of the usefulness of various EIES features or capabilities, they are ranked near the top at all levels of experience. In addition, there was a question on the follow-up questionnaire that provided for open-ended comments about user consultants. The question read as follows:

> Have you ever asked a user consultant for help?
>
> No
>
> Yes (Please describe whether this was helpful, satisfactory, courteous, or whatever.)

Unfortunately, the question was biased toward a positive response. Most users did report contact with a user consultant (82 of 108 responding answered "yes"). Of these 82, 67 made favorable comments—but most took the easy way out of simply circling one or more of the adjectives, such as "helpful," "satisfactory," or "courteous." Fuller responses are quoted in Table 5.2. Though there are a few cases of reported nonresponses from the user consultants, most of the comments are enthusiastic. Typical are "Excellent and friendly," "Fantastic!," and "They are essential."

If one were forced to live with a single source of help, then the human user consultants appear to be the best choice. However, without the online aids to serve as a resource for many of the routine questions, they would probably soon become overloaded (and not so prompt and friendly), like a housewife with too many children and not enough appliances.

User consultants appear to be a vital element in system acceptance. As Bair (1979:257, in Uhlig, Farber, and Bair) puts it:

> Although the best documentation and assistance may be available and frequent courses given, a continually available channel of communication with the (service providers) is necessary . . . The feedback mechanism should enable users to ask questions at any time, receive a response as fast as possible from an expert, and submit design suggestions which may eventually be implemented.

Reporting the results of another case history of office automation, Open Systems (1981:7) concludes that to achieve high acceptance and participation rates, "you have to do a lot of 'hand holding' initially—like 24 hours of training (and encouragement) per person—from an outside group specializing in social psychology." The EIES user consultants are not necessarily "social psychologists" (although two happened to be sociologists), but they are users themselves, not programmers. They speak the users' language, and provide a highly valued service.

DEALING WITH THE PAPER MONSTER

The massive amounts of paper generated by a medium that is supposed to be the precursor of a "paperless society" is the subject of much joking and of genuine distress. In the public conference on "Impacts," for instance, there are mentions of having to buy more and larger waste baskets and of taking out the garbage more frequently. The long rolls of thermal paper on the portable terminals provided to many users are especially difficult to store, since they are not perforated and do not easily fold or divide into pages.

As shown in Table 5.3, users vary in how they handle their printouts. Some develop complex indexing and filing systems, complete with color coding. A few go so far as to keep written logs of all messages sent, dates

TABLE 5 · 2 ***Sample Comments about the User Consultants***

"Excellent and friendly"
"Prompt!"
"Great"
"Very useful"
"Very prompt and useful"
"Very satisfactory, very courteous, very enjoyable!!!"
"Unavailable in most cases"
"Not too helpful—merely repeated what I already knew"
"Fantastic!"
"Some consultants were helpful; others were not."
"Nice people"
"Courteous, prompt, usually but not always"
"They're great—you know that!"
"Very helpful, quite courteous. They are essential."
"9 times out of 10 the response is prompt, helpful, courteous, and friendly. Occasionally a request seems to be ignored."
"Yes—but—a bit more 'kindergarten' approach needed"
"Very helpful, satisfactory, and courteous. I am very impressed by the services provided by these people."
"They were helpful and courteous and answered my question quickly."

TABLE 5 · 3 ***Disposition of Printouts, by Group***

	30	35	40	45	54	Total
Throw all out	0%	8%	0%	0%	13%	4%
Keep them all	9%	32%	10%	24%	—	18%
Save selective entries in single file	32%	12%	31%	19%	25%	23%
Save selective entries in separate files	32%	28%	45%	38%	25%	34%
Use a CRT	4%	12%	7%	14%	38%	15%
Other	23%	8%	7%	14%	38%	15%
Total responding	22	25	29	21	8	109

Chi square = 34.3, p = .10
(4 group 50 responses omitted from above)
Source: First Follow up
Question: "What do you do with the printouts of material from EIES?"

of confirmations, etc. Only a few throw away the printouts. The modal method is to establish categories by conference or group number and to file hard copies within these categories, thus simplifying retrieval and review.

It is likely that any computer-mediated communication system in which a user is dealing with multiple conferences and documents will stimulate the squirrel-like behavior of keeping paper printouts rather than throwing them away. While a CRT is fine for dealing with short messages, it is very tedious and unsatisfying for reading long documents and conference transcripts. As John Senders puts it (in Sheridan, et al, 1981:78),

> The question is not yet answered as to whether the desire to have hard copy for perusal at leisure is merely a consequence of practice with printed materials or whether it is a natural and more or less immutable characteristic of human beings.

Thus, employers must anticipate that users will keep printouts and could benefit from suggested techniques for systematically dealing with the monstrous piles of paper that will be generated.

REACTIONS TO THE INTERFACES AND RELATED FEATURES

A design decision was made to provide a variety of alternative interfaces. The theory was that they would form a progression. Most users would start out with the long menu, which requires no knowledge of the system. Users would then progress to the short menu. From there they would begin utilizing answer-aheads and commands until, finally, they would frequently use their own defined commands or strings of operations.

The data show that there is such a progression (see Table 5.4). However, they also show that there is a great deal of individual variation in interface preferences and patterns of use. Although it is true that the long menu is the preferred interface for new users and becomes less frequently used the more experience a person has, its use never stops altogether. Among those with 50 or more cumulative hours online, 41% report that they "sometimes" use the long menu. Apparently, they turn it on when they use new or unfamiliar parts of the system or when they

TABLE 5 · 4 ***Use of Alternative Interfaces, by Time Online: Percentage Using Interface "Frequently" or "Often"***

	Hours online				
	<20	20–49	50–99	100+	All
Long menu	45%	36%	0	0	33%
Short menu	41%	40%	69%	50%	44%
Answer ahead	20%	44%	58%	75%	39%
Commands	34%	71%	74%	75%	48%
String variables	0	3%	43%	25%	10%
N responding	41	35	12	8	96

Source: Follow Up Questionnaires

have been away from the system for a while and need to refresh their memories. (This figure does not appear in Table 5.4, which shows only the frequency of the "frequently" and "often" responses.) Thus, though there is a tendency for the predicted progression, the interface cannot be automatically changed at a certain point in time. After experience is gained, commands are the most frequently used interface, but the others are used either habitually or occasionally by a large proportion of the system's members.

USER SUPPORT FOR LEARNING MENUS FIRST

Users who have previously used command-driven systems are sometimes impatient with the menu as an introductory interface. However, the majority of EIES users support the design decision to teach menus first (Table 5.5). Using the menu seems to have the cognitive effect of helping the user to develop a mental map of the structure of the system. When the user understands the structure of the system and the relationship among components, the more active command mode can be used to move around the system at will. Support for the menu as a beginning interface grows stronger the more time a person spends online.

FORCED DELIVERY OF MESSAGES

A somewhat more controversial aspect of the EIES design is that, although users may postpone delivery of messages, such undelivered messages will remain in the queue, and the user will be frequently notified

TABLE 5 · 5 ***Preference for Teaching of Menus or Commands First, by Time Online***

	Menus First	Commands First	Other	N
5–19 hours	52%	38%	10%	29
20–49 hours	74%	22%	11%	29
50–99 hours	83%	8%	8%	12
100+ hours	88%	12%	0%	8
Total	70%	24%	6%	78

Source: Follow Up Questionnaire
Question: Do you now think it is a good idea or a poor idea to introduce the new user to the system through menus and to provide equivalent commands for those who prefer them?

of their pending status until they are accepted. Some users wish to be able to reject the delivery of messages without printing them out, perhaps on the basis of author or keys. Forced delivery is not made of items in conferences or notebooks, where members are free to read a header only, the full text, or nothing.

The designer's point of view is that confidence that a message sent will actually be delivered is more important than the temporary inconvenience of a recipient. Furthermore, if a person is sending overly wordy or irrelevant messages, other group members should let him or her know, rather than surreptitiously refusing delivery of further messages from the person. A particularly sticky design argument is what to do about confirmations if rejection of messages were indeed permitted. Since delivery of all messages is normally confirmed, should a comparable rejection notice be returned to an author? What would it look like? Confirmations normally state, "Mxxx received by (name) at (date and time)." Should messages rejected result in a comparable confirmation of "Mxxx rejected by (name) at (date and time)." What might this do to social relationships on the system?

One of the most popular design alternatives suggested is to make acceptance of group messages (but not of private messages) optional, in which case authors could at any point check a confirmation list if they want to know who has actually read a message. The other is to permit rejection of any message with some sort of notification. For all users, the modal preference is support of the current design, with forced (eventual) acceptance of all messages. This is endorsed by half the members responding overall, and the support of the design decision increased with experience (see Table 5.6). The second most popular option, endorsed by a quarter of users, is to allow rejection of any message, with notification to the author. Many people suggest some kinder term than "rejected" or "refused," such as "NAME has been notified of pending M###." And about 4% suggest some other alternative altogether. Thus, there is no one solution to this problem that will satisfy everyone, but the forced delivery of at least private messages is generally endorsed.

LIKED AND DISLIKED FEATURES OF THE SYSTEM

The post-use questionnaire included open-ended questions on those aspects of the system that are considered to be most valuable and useful vs. those most useless, distracting, or in need of improvement.

TABLE 5 · 6 ***Percentage of Users Favoring the Requirement that All Messages Must Be Accepted by Addressees, by Time Online***

Cumulative Hours	%
<20 hours	43%
20–49 hours	51%
50–99 hours	58%
100+ hours	71%
All (N = 103)	50%

Source: Follow Up Questionnaire
Question: In EIES, you do not have the choice of permanently refusing to accept a private or group message. Which of the following would you prefer?
Require acceptance of all messages, as at present.
Require acceptance of private messages only.
Allow rejection of any message, with "message refused by ###" returned to the sender.
Comments?

EIES FAVORITES

Table 5.7 shows a selection of answers evoked by the open-ended question on the "most valuable features of EIES." (For a complete list of responses to this and other open-ended questions, see Hiltz, 1981).

Note that many members do not mention specific features. Instead they cite general characteristics and advantages of the medium, such as the fact that users "self-organize" information and experience the intellectual stimulation of a wide range of contacts. Among those who name specific features as the most valuable aspect of EIES, messages and conferences are most frequently singled out, but text editing and joint notebooks are also frequent nominees for "Best Feature." In addition, many relatively "minor" aspects of the design are singled out as somebody's very favorite feature, such as "+link" (real-time interchange of single lines of text), the "Paper Fair" (an unrefereed online journal), the multiple interfaces, the directory, and even the short but friendly "Welcome" that is the way the EIES computer responds when first dialed. This diversity underscores the conclusion reached in the analysis of "The Evolution of User Behavior" that there is no single design based on a small number of features that will satisfy the experienced, sophisticated user. Users begin to be gourmets, appreciating the subtleties of the choices and variations that can be selected from according to preference.

A third group of responses focuses neither on general medium characteristics nor on specific features, but on specific benefits derived from using EIES. This includes decreased need to travel; the ability to obtain such things as annotated bibliographies of recently published material, contributed by group members; and the opportunity to interact with well-known scholars. (The graduate student who wrote this noted that such collegial contact with well-known scholars at other institutions would not otherwise have been possible).

COMPLAINTS ABOUT SPECIFIC FEATURES AND CHARACTERISTICS OF THE MEDIUM

Most of the complaints about "useless" or "distracting" features are so specific to EIES at a particular time that they have been omitted from this account. The length of the complaint lists and the specific nature of the suggested improvements vary markedly by group. Groups 30 and 45 have very few members who list anything as useless or distracting. On the other hand, Groups 40 and 54 (especially relative to the small number of post-use returns from group 54) have many nominees for "Worst Feature." What is happening in many cases is that annoyance at the social (or more precisely, asocial) behavior of group members is frequently responded to as the worst "feature of the system," when of course this is not software at all, but rather disappointment with some of the interaction online.

Group 35 has several complaints about the quality of the content of the communications contributed by its own members ("junk messages," "off-the-wall comments," "making cute remarks"). Similarly, Group 40 is marked by the number of complaints about group messages that are voluminous, unnecessary, or of little general interest. This group had the largest number of "Season's Greetings" consisting of graphics and text exchanged as group messages. Apparently, some of the members strongly disliked this particular form of electronic art used as social amenity, particularly if they were off-line during the holiday season and had to sit through a dozen or so Christmas trees printing out in February. As they themselves suggest, one solution would be that group messages should have a self-destruct date. That is, when they are entered, the sender should be asked the last date on which the message should be delivered, since most group messages refer to subjects of interest for only a limited time.

Subsequently, features were developed on EIES to allow users to send a short group message that contains embedded within it a much lengthier discussion for those who are interested in reading it. (There are actually two methods for doing this, one suited for one page of

TABLE 5 · 7 ***A List of the Most Valuable Features of EIES (Paraphrases of Selected Responses to an Open-Ended Query: Post-use Questionnaire)***

I especially liked the immediacy of communication and the diversity of discussions in which I could participate actively or passively. It was fun and intellectually stimulating to be part of EIES.

The asynchronous mode of communication is the most valuable feature; it allows both for delayed responses and for the delivery of messages whenever and as soon as the addressee returns to his/her terminal.

Group conferences: The sharing of ideas is valuable.

There are many levels of interface.

EIES is really designed for humans! One does feel free on EIES, not constrained by the computer. It allows the user to utilize "natural" information processes and strategies. Getting information from people is pleasant and efficient. Information is not preorganized, like in data bases; it is "self-organized" by the users.

The directory and the search/retrieval processes are, in general, quite informative and easy to use. (And the "Welcome" header for new members is great as a first introduction to the people using the system.)

I feel no pressure to say anything in conferences. I've learned more by listening more.

Getting annotated reading suggestions is a great learning tool.

The speed of communication is a big plus.

There is availability of the entire history of a conference. Messages can also be private, and the personal message exchange is very useful.

The ability to send group messages.

The ability to send instantaneous private messages and to participate simultaneously in group conferences.

One can quickly tap both special and varied information.

One gets a real feeling of living in a network society.

There are many time-saving system commands to do things directly.

TABLE 5 · 7 ***(cont.)***

There are many editing features built intrinsically into the system.

I was able to interact with some well-known scholars (and with the advantage of instant interchange!)

I especially liked the new "paper section," c1010

There are a large number of interesting and active people; there is always mail or a new conference item of some interest.

It is very easy to sign off.

There is the ability to reach anyone on the entire EIES system.

---

material, the other for making many pages available on request). Recent observation shows many fewer group messages being sent and the frequent use of the mechanisms for making a short announcement followed by material that does not print unless selected by the recipient.

Group 54 is the only group with complaints about the basic system design. Many of these group members were used to working on very sophisticated, high-baud-rate systems at their own universities. They should probably not have used EIES at a low baud rate, but should have used micros as terminals so that they could edit with a familiar editor and scan material at the high-baud-rate to which they were accustomed.

However, Group 54's complaints probably also reflect an insufficient level and frequency of use to maintain facility with the system. As Bair (1979, in Uhlig, Farber, and Bair) points out, unless people use a system at least every few days they keep forgetting what they learned, and the system always seems difficult and arbitrary. Group 54 never got any successful conference or activity going online, with the exception of a period after Three Mile Island when the nuclear accident inspired considerable analysis of the reasons for the disaster.

SUBJECTIVE SATISFACTION SCALES

For what kinds of communications functions is EIES most suitable? How does it "feel" to use this system? And how do reactions to EIES compare to those for other systems with different features and interfaces?

TABLE 5 · 8 ***Experiences while Communicating via EIES and Similar Systems***

	Always	Almost Always	Some-times	Almost Never	Never	EIES Mean	TELEMAIL Mean	PLANET Mean
Distracted by the mechanics of the system	5%	16%	49%	23%	7%	3.1	3.3	3.2
Constrained in the types of contributions you could make	4%	17%	44%	28%	8%	3.2	3.1	3.6
Overloaded with information	4%	18%	55%	16%	6%	3.0	4.1	3.6
Able to express your views	24%	47%	24%	5%	0	2.1	2.1	2.0
Able to get an impression of personal contact with other participants	8%	35%	46%	6%	5%	2.6	2.2	2.6

Source: EIES Follow-Up Questionnaire, N = 110
TELEMAIL Follow-Up Questionnaire, N = 22
PLANET means computed from raw data reported on p. 182 of Vallee et al., 1978.
(Question: Thinking back over your experiences with the system, how frequently have you felt . . .)

EXPERIENCES WHILE COMMUNICATING OVER EIES

Table 5.8 shows the frequency with which users report various experiences or feelings while using EIES. For example, most users "sometimes" feel distracted by the mechanics of the system intruding upon the natural flow of communication. There is a tendency for this feeling to decrease with more time online, but the relationship is not statistically significant (gamma = .17, p = .24).

For feeling "overloaded with information," "sometimes" is also the modal answer. The frequency of feeling an overload appears to peak in the middle ranges of use; 31% of those who had logged 20–49 hours online report "almost always" experiencing information overload, whereas all of those who have logged 100 hours or more report the overload experience to occur only "sometimes." Apparently, experienced users develop effective ways of coping with what may initially seem to be an "overload" of communication.

Being "able to express your views" is generally reported to occur "almost always;" among those with 50 hours or more online, the responses are all in the "always" or "almost always" category. (The correlation with hours online as measured by gamma is −.20, significant at the .05 level).

In terms of feeling "constrained in the types of contributions you could make," "sometimes" is also the modal answer. Finally, being "able to get an impression of personal contact with other participants" tends to occur "sometimes" for those with less than 50 hours online and "always" or "almost always" for those with more than 50 hours online (correlation with hours on line, as measured by gamma, is −.47, significant at the .02 level).

Comparable Data for PLANET and TELEMAIL. Most of the averages for the frequency with which users have these feelings are very close (see the last columns of Table 5.8). Users of all three systems tend to "sometimes" feel distracted by the mechanics, to "sometimes" feel constrained, "almost always" able to express their views, and, somewhere in the "sometimes" to "almost always" range, able to get an impression of personal contact with others. The only difference is that the users of the mail system less frequently feel overloaded with information than do the users of the two conferencing systems, who sometimes find many items waiting for them in a large group conference.

OVERALL SATISFACTION RATINGS

Overall ratings of EIES as a communication-information system are fairly positive, but not "perfect," in terms of users' subjective responses to a number of satisfaction scales. Subjective reactions are weakly correlated with total amount of use of the system, with higher satisfaction levels expressed by those with more time online.

Table 5.9 shows that users tend to rate the system as good overall by the three-month follow-up and also as stimulating rather than boring, productive, fun, friendly, and easy to use. There are three dimensions on which a quarter to a third of the respondents give a negative rating: that the system seems to be frustrating at times, time wasting, and intrusive.

These subjective satisfaction scales are the most general assessments of EIES that we have. They will be used as the basis for a more detailed analysis of subjective satisfaction factors and their determinants, which will be presented at the end of the chapter.

Subjective Satisfaction Ratings and Time Online. The subjective ratings of EIES do tend to be positively related to accumulated hours online at the time the questions were answered. However, most of the relationships are weak and statistically insignificant. The overall rating of the system (EIES is extremely good-extremely bad) is significantly related to time online (Chi-square = 32.6, p = .04, gamma = −.45). The only other scales showing a significant relationship are personal-impersonal (gamma = −.24, p = .01) and time saving-time wasting (gamma = −.28, p = .05).

Comparable Data for Theory Net on TELEMAIL. The last column of Table 5.9 shows the comparable means for overall subjective satisfaction ratings for TELEMAIL. There are many similarities, such as almost exactly the same average for the overall rating of the systems as "extremely good" to "extremely bad." The means are exactly the same for "easy" to "difficult." However, there are also some interesting differences. EIES is seen as more friendly, more stimulating, and more fun. However, it is also seen as more time wasting and demanding, probably because of the much larger volume of activities online.

THE DACOM SCALES

For what communications functions and processes is a system like EIES most satisfactory? The DACOM scales (Description And Classification of Meetings), designed originally by the Communications Studies Group in Great Britain (see Short, Williams, and Christie, 1976), have been used to measure users' perceptions of a variety of systems and media. For this study, scales were administered both at follow-up and at post-use. There were very high correlations between the measures at the two points in time. The post-use data were chosen for Table 5.10. For each of a number of communications functions, users are asked in these scales to rate a medium from "1" (completely satisfactory) to "7" (completely unsatisfactory).

EIES is seen as most satisfactory for emotionally neutral task-oriented functions: giving or receiving information, exchanging opinions, generating ideas, giving or receiving orders. It is also seen as satisfactory by most people for social-emotional tasks such as getting to know someone and expressing positive and negative emotions. It is perceived as least satisfactory for functions related to conflict and negotiation: problem solving, bargaining, persuasion, resolving disagreements. For these last tasks, the ratings cluster in the neutral (3–5) range rather than in the positive (1–3) range characteristic for other functions.

Two groups using the system (JEDEC and a medical standards group called MRFIT) reported that a characteristic communication mix was to use the system for routine communication and to resort to other modes, such as face-to-face meetings or the telephone, when conflict arose. Whether special structures can be incorporated into computerized conferencing systems to support conflict resolution is a research question that is now being pursued. Without such special structures, it is evident that user groups find the medium lacking for conflict resolution tasks.

Group Differences. Most of the DACOM scales do not show significant differences associated with the user group. However, some do. Using the system for persuasion is most highly rated by members of Group 40 and received the most unsatisfactory ratings from Group 45. Resolving disagreements, significant at only the .09 level, showed a similar pattern. This is to be expected, since they are similar functions.

For "getting to know someone," Group 30 is the most positive, followed by Group 40, and Group 45 is again the most negative. For giving and receiving orders, on the other hand, Group 45 is more split than the others, and Group 54 is the most decidedly neutral.

None of the other scale items show differences among groups that are significant at the .10 level or above. The differences that do occur indicate that the specific experiences of the group do have some effect upon ratings of the degree to which the system in the abstract is suitable for some communications purposes. In other words, as we have pointed out in Chapter 3, the nature of the application and social context that is bound up with "group" has a mediating effect upon perceptions of the system itself.

Comparable DACOM Scale Results for PLANET and TELEMAIL. The DACOM ratings of the extent to which MACC-TELEMAIL, EIES, and PLANET were satisfactory for specific communications functions (Table 5.11) yielded similar results for most items, with the exception of "getting to know someone." For all three systems, giving or receiving information and exchanging opinions were the tasks for which the highest degree of satisfaction was reported; bargaining and persuasion were among the

TABLE 5 · 9 ***Overall Reactions to the EIES Mode of Communication***

1	2	3	4	5	6	7	EIES Mean	TELEMAIL Mean
Extremely good						Extremely bad		
12%	78%	30%	3%	10%	4%	0	2.8	2.9
1	2	3	4	5	6	7		
Stimu-lating						Boring		
20%	37%	27%	7%	8%	1%	0	2.5	3.9
1	2	3	4	5	6	7		
Productive						Unproductive		
6%	22%	36%	16%	14%	4%	3%	3.3	3.1
1	2	3	4	5	6	7		
Great fun						Unpleasant work		
15%	36%	22%	14%	12%	1%	0	2.7	3.9

1	2	3	4	5	6	7		
Time saving						Time wasting		
10%	15%	19%	24%	22%	6%	4%	3.7	2.3
1	2	3	4	5	6	7		
Not frustrating						Frustrating		
8%	11%	22%	22%	23%	9%	4%	3.9	3.9
1	2	3	4	5	6	7		
Friendly						Impersonal		
16%	42%	28%	14%	6%	3%	0	2.7	3.9
1	2	3	4	5	6	7		
Easy						Difficult		
16%	28%	22%	18%	13%	3%	0	2.9	2.9
1	2	3	4	5	6	7		
Not demanding or intrusive						Very demanding or intrusive		
14%	16%	20%	22%	24%	3%	1%	3.4	1.4

Source: EIES Long follow-ups, N = 111
MACC-TELEMAIL Follow-up Questionnaires (N = 22)

TABLE 5 · 10 ***CSG DACOM Scales: Extent to Which EIES Is Satisfactory for Various Communications Tasks***

	Completely Satisfactory 1	2	3	4	5	Completely Unsatisfactory 6	7	Means	Gamma	p*
Giving or receiving information	25%	41	14	10	7	4	0	2.4	.17	.70
Problem Solving	3%	15	19	28	23	7	4	3.9	.15	.22
Bargaining	6%	9	16	30	20	9	9	4.1	.16	.65
Generating ideas	15%	30	33	11	7	1	3	2.8	.29	.46
Persuasion	4%	5	29	20	19	15	8	4.2	.23	.02
Resolving disagreements	5%	7	28	23	16	14	7	4.1	.18	.11
Getting to know someone	5%	29	35	14	7	7	4	3.3	.21	.26

Giving or receiving orders	10%	34	15	22	8	5	6	3.2	.08	.65
Exchanging opinions	25%	42	20	6	5	1	2	2.3	.18	.26
Expressing positive emotions	7%	25	33	16	8	4	6	3.3	.04	.20
Expressing negative emotions	7%	22	22	22	16	5	5	3.5	.17	.66
Sociable relaxation	2%	21	27	21	11	7	10	3.9	.17	.42

Source: Post-Use Questionnaire, N = 102

**Gamma is a measure of association with hours of use.*

p (probability) is the level of significance of this association, based on Chi-square.

TABLE 5 · 11 ***DACOM Scale Measures—MACC-TELEMAIL, EIES, and PLANET***

Means

Function	TELEMAIL	EIES	PLANET
Giving or receiving information	2.0	2.4	2.1
Problem solving	4.0	3.9	3.4
Bargaining	4.4	4.1	4.2
Generating ideas	3.8	2.8	2.6
Persuasion	4.3	4.2	4.6
Resolving disagreements	3.5	4.1	4.3
Getting to know someone	4.8	3.3	4.5
Giving or receiving orders	3.2	3.2	2.4
Exchanging opinions	1.9	2.3	2.1

Sources: Theory Net Follow-Up Questionnaires (approximately 18 months; N = 22)
EIES Post-Use Questionnaires (approximately 18 months; N = 102)
PLANET: Computed means to nearest .1 from raw data included on p. 183 of Vallee et al., 1978. Scale reversal used to obtain comparable values.
Questions: How satisfactory do you think the system is for the following activities? (1 = completely satisfactory, 7 = completely unsatisfactory)

least satisfactory for computer-mediated communication. Using a criterion of more than a one point difference between means, the only clear difference is in "getting to know someone"; for this, EIES received higher ratings. This can probably be attributed to differences in design, such as the presence of a public directory in EIES and the group vs. individual orientation of conferences as compared to messaging. Another apparent difference is that the TELEMAIL group does not seem to have as much difficulty with resolving disagreements online. This may be because group debates are seldom held via messages, as compared to conferences, which are often set up specifically to find and discuss differences of opinion. It is also partially attributable to the fact that the Theory Net group is not in a scientific community that is undergoing a lot of disagreements.

PERCEIVED BALANCE BETWEEN EFFORT AND BENEFITS

One measure of satisfaction with the EIES experience lies in the feeling by participants that they received benefits at least equal to the effort expended. The majority of participants, as shown in Table 5.12, do feel that they received as much or more than they contributed to their group(s). The most active participants (100 or more hours online) are most likely to perceive a balance between their contributions and the amount and quality of information received from others. Surprisingly, participants at the lowest levels of activity, who are most likely to receive much more than they contribute to the system, do not always perceive this to be the case. At intermediate levels of use (50–99 hours online by the end of the trials), there is the greatest probability that participants will feel that they are contributing more than they are receiving in return.

SYSTEM AS USELESS TO REVOLUTIONARY: TELEMEIL, EIES, AND NLS

An item designed by Edwards for her NLS evaluation was used for the EIES and TELEMAIL studies in order to obtain comparable measures of feelings about the usefulness of the systems and the extent to which they were potentially "revolutionary" (see Table 5.13). Remembering that dissatisfied or low-level users were least likely to complete the questionnaires, it is not surprising that, for all systems, responses are generally more positive than the "neutral" point and that users are likely to feel that their system has at least "certain worthwhile uses." The only clear difference seems to be in the extent to which users feel that the system is "revolutionizing" their work and communications. This is not at all as frequent an evaluation for the simple mail system as for the more complex systems designed to support a wider variety of communications and work functions.

ATTITUDES OF OTHER HOUSEHOLD MEMBERS

A communications system like EIES potentially generates reactions not just from direct users. Others observe interaction with the system and form opinions about whether this online activity adds to, detracts from, or is neutral in terms of its effects on their (off-line) relationship. The most important of the potential groups on which there may be a secondary impact is the family, particularly if the network member uses a terminal at home, and particularly if he or she ties up the only phone line.

Many EIES users do not take their terminal home or talk to their families about their work; their families or living partners are oblivious

TABLE 5 · 12 ***Balance between Contributions Made and Information Received, By Time Online***

	Contributed Sig. More	Contributed More	Equal	Received More	Received Sig. More	N
1–19	4%	7	37	37	15	27
20–49	6%	9	36	21	27	33
50–99	12%	28	28	12	20	25
100+	0	10	58	16	16	19
All	6%	14	38	22	20	104

Source: Post-Use Questionnaire
Chi-square = 16.5, $p = .17$, gamma = .14
Question: Comparing my contributions or effort put into EIES with the amount of information received, I feel that I have: contributed significantly more than I have received, contributed more than I have received, contributed as much as I have received (equal), received more, received significantly more than I have contributed.

TABLE 5 · 13 ***Overall Ratings of Systems as Useless to Revolutionary: EIES, TELEMAIL, and NLS***

	EIES	TELEMAIL	NLS
I think it is useless and should be discontinued.	0	0	1%
I think it has its uses for others, but not for me.	4%	0	1%
I am skeptical but am giving it a try.	8%	0	5%
I am basically indifferent or neutral.	0	0	3%
I think that it has certain worthwhile uses for me.	41%	47%	22%
I think it is very useful in many respects.	31%	47%	44%
I think it is revolutionizing my work/ communications processes.	17%	5%	23%
Total	100%	100%	100%
N	107	19	94

Sources:
EIES: Follow-Up Questionnaire
TELEMAIL: Follow-Up Questionnaire
NLS: Post-Use Questionnaire (Edwards, 1977, p. 105)
Question: Which statement best describes your present reaction to . . .

to it. For those who do take it home, reported reactions vary from great curiosity and enthusiasm to hostility and resentment (see Table 5.14). Reactions of interest, curiosity, and support are much more frequently reported than are negative reactions.

A very lively debate on the impacts of EIES on family life occured in the public conference on "Impacts." Opinions ranged from the point of view that "CC will worsen the detrimental strain that TV and other relatively modern technical developments have put on family bonds" to

TABLE 5 · 14 ***Sample Responses of Other Family Members or Friends***

"Seems like a fun thing that I am doing, but it is no big deal to them."

"Curious fascination to irritation (when I bring the terminal home)."

"Huh? They couldn't care less."

"Very interested."

"Enthusiastic, interested, envious in friendly fashion; and they learn things from EIES."

"My wife is moderately interested. My children are enthusiastic."

"Interested. Look for future developments in this technology."

"They dislike my keeping the phone busy too frequently and too long each time."

"Curious skepticism."

"My wife likes it a lot. My wife checks the messages and 'talks' with the systems people."

"Kid plays 'animal' on visits."

"Don't know or care."

"Think it is a fun toy. Are annoyed at tying up the telephone. Are interested in messages that they understand."

"They hardly know."

"That it's great and should be expanded to all areas of communication."

"Oblivious."

TABLE 5 · 14 ***(cont.)***

"Tolerant; not excited at all."

"Positive."

"Children neutral. Wife negative."

"I have been forced into mainly working on EIES after 5:00 pm because of telephone rates. My occasional latenesses in returning home annoy my wife."

"My husband is interested and a bit envious. My children are too young to understand what it is all about, but accept it matter-of-factly."

"Indifferent."

"Mildly interested."

"They find it terribly exciting; 'A giant intellectual C.B.,' as one of them described it."

"Impressed."

"Amusement and amazement."

"Interested."

"Encouraging."

"They have no attitudes whatsoever toward it."

"They know nothing about it. It's my dark secret."

"Yet ANOTHER activity to distract me from family life! But generally supportive!"

"Respect and admiration."

TABLE 5 · 14 ***Sample Responses of Other Family Members or Friends, (cont.)***

"Wife enjoys it, finds it interesting and amusing."

"They are disappointed that, unlike other computer systems I interact with, EIES has no provisions for interstellar combat and similar diversions."

"That damn computer."

"Wife: indifferent. Children: somewhat curious."

"Supportive."

"They think it is interesting . . . like a toy."

"Amused—sometimes annoyed."

"A distraction, but they accept it as important."

"Love it."

"Intrigued"

"Enjoyable."

"My wife is excited about the idea and the system."

"Moderately interested."

"Very positive—after we got an extra telephone line for the terminal."

"Positive except when 1) paper accumulates throughout the house, or 2) I become frustrated when system is slow or I have difficulty accomplishing what I intend to do."

"Between EIES and my home computer they sometimes wonder who that strange man is in the study."

"Supportive, interested, excited."

TABLE 5 · 14 ***(cont.)***

"My children are not involved in and not aware of EIES. My wife knows about it and thinks it's great."

"Wife is a user."

"Wife has mild interest when I take it home."

"Enthusiastic—amazed."

"I don't use it at home. If I did, it might compete with family activities."

"My wife is excited about the idea and the system."

---

the assertion that it can strenthen the family by, for instance, allowing spouses separated by travel to remain in contact or permitting parents to work at home rather than leaving their children to go to an office. The intensity of many of the comments on the "pros" and "cons" of having a terminal in the home indicate that the reactions of other household members to CC as well as those of primary users should be included in future studies of acceptance of the medium and its impacts.

MULTI-VARIATE ANALYSIS: FACTOR ANALYSIS AND STEPWISE MULTIPLE REGRESSION

An attempt at multi-variate analysis was hindered by the number of cases available when using many variables from the pre-use and follow-up questionnaires: if there is no answer available on one of the variables used, then the case is eliminated from the analysis. The more variables in the multi-variate analysis, the more the number of cases shrinks. Nevertheless, we found some interesting clustering of subjective satisfaction measures, and some important determinants.

The items shown in Table 5.15 were subjected to one of the most widely used approaches to factor analysis, the "PA2" approach (See Kim, 1970) with VARIMAX rotation. (This is the "normal" or "default" type of factor analysis in SPSS). The factor analysis showed how dimensions of subjective satisfaction cluster together, and could be combined to derive underlying factors which several of the individual questions have in common.

Two underlying factors were identified. The first seems to correspond to "input frustration" or difficulty. It includes the items on the extent to which the system is frustrating or hard. The second seems to be "output payoff" satisfaction, and includes the items on whether the system is good, stimulating, productive and fun.[1] Three questions were right in the middle of both factors, which makes sense, because they correspond to a balance between input difficulty and output payoff.

Having identified the INPUT and PAYOFF factors, index scores were constructed by adding together the scores for the component questions. These were used in two stepwise multiple regressions, with several predictors entered in order to discover which ones are the most powerful determinants of these dimensions of subjective satisfaction.[2] (See Chapter 2 for an explanation of the nature and purpose of stepwise multiple regressions).

The variables used in the analysis are:

KNOWN = Number of group members known before system use

ESTUSE = estimated number of hours of use per week, before using EIES

NUMBER = Number of persons with whom the user felt in active communication on EIES at follow-up

EIES MET = Number of these persons "met" on EIES

[1]The "varimax rotated factor matrix" is shown below, divided into those variables which were combined into a "payoff factor" index, those which combined into an "input difficulty" index, and those which were omitted because they do not load clearly on one factor. The scores are the regression weights of the common factors. (See Table 5.12 for a complete list of the words used on the individual semantic differential scales).

Variable	*Factor 1*	*Factor 2*	*Variable*	*Factor 1*	*Factor 2*
PAYOFF factor			(Related to Neither or Both—not used)		
Good	.73	.39	Time-saving	.59	.44
Stimulating	.86	.16	Friendly	.46	.32
Productive	.78	.33	Demanding	.27	.23
Fun	.68	.19			
Input Difficulty Factor (INPUT)					
Frustrating	.30	.64			
Easy	.16	.72			

[2]The variables to enter into the multiple regression were selected by first computing bivariate Pearson's correlations and significance levels for the relationship between the two indexes and several possible predictors. Group, previous experience with terminals, and satisfaction with the group leader were eliminated because they did not yield significant correlations.

TABLE 5 · 15 ***Stepwise Multiple Regression: Determinants of PAYOFF Satisfaction Factors***

Correlation Matrix

	KNOWN	ESTUSE	NUMBER	EIESMET	HRSUSE
PAYOFF	.14	.33	.22	.26	.24
KNOWN		.20	.47	.13	.41
ESTUSE			.16	.21	.46
NUMBER				.44	.38
HRSUSE					.40

N of cases = 44

Stepwise Multiple Regression

Factor	Mult R	R Square	Beta	P
ESTUSE	.33	.11	−.28	.01
EIESMET	.38	.15	−.15	.05
NUMBER	.40	.16	−.10	.10

Note: See text for definition of variables

HRSUSE = Number of hours spent online at time of follow-up

The stepwise analysis for "PAYOFF" (see Table 5.15) shows that the most important determinant is the attitude toward the system before using it, as indicated by estimated hours of use. Once again, we come up with the finding that users somehow knew before communicating on EIES how much they would like the system and how much they were likely to use it and benefit from it. As shown in Chapter 2, the strongest observed correlate of preuse estimates of EIES use is the number of group members known. The variable entered at the second step of the "PAYOFF" analysis, which significantly improves the prediction, is the number of persons "met" on EIES. A third variable which improves the prediction somewhat (significant at the .10 level but not the .05 level) is the number of persons with whom one is communicating on EIES. In sum, our most important determinants of satisfaction with what one is getting out of EIES are measures of social connectivity. Once these variables are taken into account, time online has no independent effect.

None of the variables were significantly related to INPUT satisfaction. The strongest predictor is the number of members known before using EIES. If one knows many other group members, one is not likely to feel that using the system is frustrating or difficult. The second most powerful predictor is the number of new persons met on EIES. Even though these findings are not based on enough cases to yield statistical significance, they are fascinating. Users's reactions to trying to use the system seem to be determined by social factors—(number of old and new communication partners). INPUT satisfaction is not related to non-social factors such as previous use of computers or computer terminals, or number of hours of experience using the system.

Just for curiosity, the variable "EIES is Not demanding or intrusive—demanding or intrusive" was correlated with the same set of predictors, since it was furthest away from either of the other factors. The most important determinant of this subjective evaluation is the group to which the user belongs (significant at the .05 level). This finding fits in with our previous observation that things most and least liked about EIES correlated with group membership.

SUMMARY, CONCLUSIONS, AND SPECULATIONS

The findings presented in this chapter have many implications for the choice of software features in a computer-mediated communication system, and also for what might be termed the "social implementation" process.

1. Although the written documentation (manual) is given generally good ratings, many users will not read through such lengthy printed material. Moreover, the standard introductory manual does not cover advanced or new features. An online explanation file, which is complete and up to date, is hundreds of pages long. Although one can search for and retrieve information on just those features of interest (like turning to the appropriate page in a printed manual), it seems to intimidate many users. Among the variety of alternative sources of "online" help and documentation provided, the most popular is the human user consultants.
2. EIES users' behavior and opinions support the design choice to provide a variety of alternative interfaces, with menus presented first.
3. Users are most likely to name as "the most valuable feature" of EIES not a specific software capability but rather general char-

acteristics or benefits of the medium, related to the people who use it, such as "diversity of discussions" or "sharing of ideas." In discussing "useless, distracting or out of place" EIES features, there are frequent complaints about slow system response time, the editor, and the difficulty of remembering the various commands and procedures for interacting with the system. However, the single most frequent category of complaint relates not to the computer system features but to the behavior or quality of performance of those with whom one is communicating: "junk mail," "cute remarks," "useless material" entered, etc.

4. One serious adaptation problem for users of this medium is "information overload." About one in five users "always" or "almost always" feels overloaded with material pouring out of the system, and the majority feel this way at least "sometimes." However, feelings of "information overload" peak at middle levels of experience, and then decrease markedly, even though the users with the most time online are objectively handling greater amounts of information. The most experienced users have learned how to cope with the rich but potentially overwhelming plethora of materials available to them online. How they prevent "information overload" at high levels of activity online is an important topic for further study.
5. When the terminal is used at home, other household members frequently develop strong positive or negative attitudes toward the system. Perhaps employees will need a new kind of counselling if the family is to successfully adjust to life in an "electronic cottage."
6. Based on the limited comparative data available, there is a great deal of similarity in user ratings of the characteristics of the four systems covered in this chapter (EIES, PLANET, NLS, MACC-TELEMAIL), despite many differences in system design. The main difference seems to be between the simple message system (MACC-TELEMAIL) and the more complex systems. The simple mail system is less "friendly," less fun, less stimulating, less useful for "generating ideas," and overall, less "revolutionary" in its impacts on users. On the other hand, it also takes much less of its users' time, is felt to be less demanding and intrusive on them, and less likely to overload them with information.
7. Multivariate analysis indicates that the most important determinants of overall subjective satisfaction with communicating on EIES are aspects of social connnectivity: how many system members one knows before signing online, how many people one actively communicates with though the system, how many valued new relationships are begun with people "met" on EIES.

Perhaps these systems are like parties. The software is like the refreshments, furnishings and decor. They can help people to enjoy themselves and communicate easily, or they can detract from the occasion. But the main determinant of whether it was a "good" party is the people there and the quality of the social interaction at the party. The party may be held in a mansion and catered by Julia Child, but if nobody talks to you, you don't like it. On the other hand, the party may be held sitting on the floor and feature beer and pretzels . . . but if all your dearest friends and most valued colleagues are there, you will have a wonderful time.

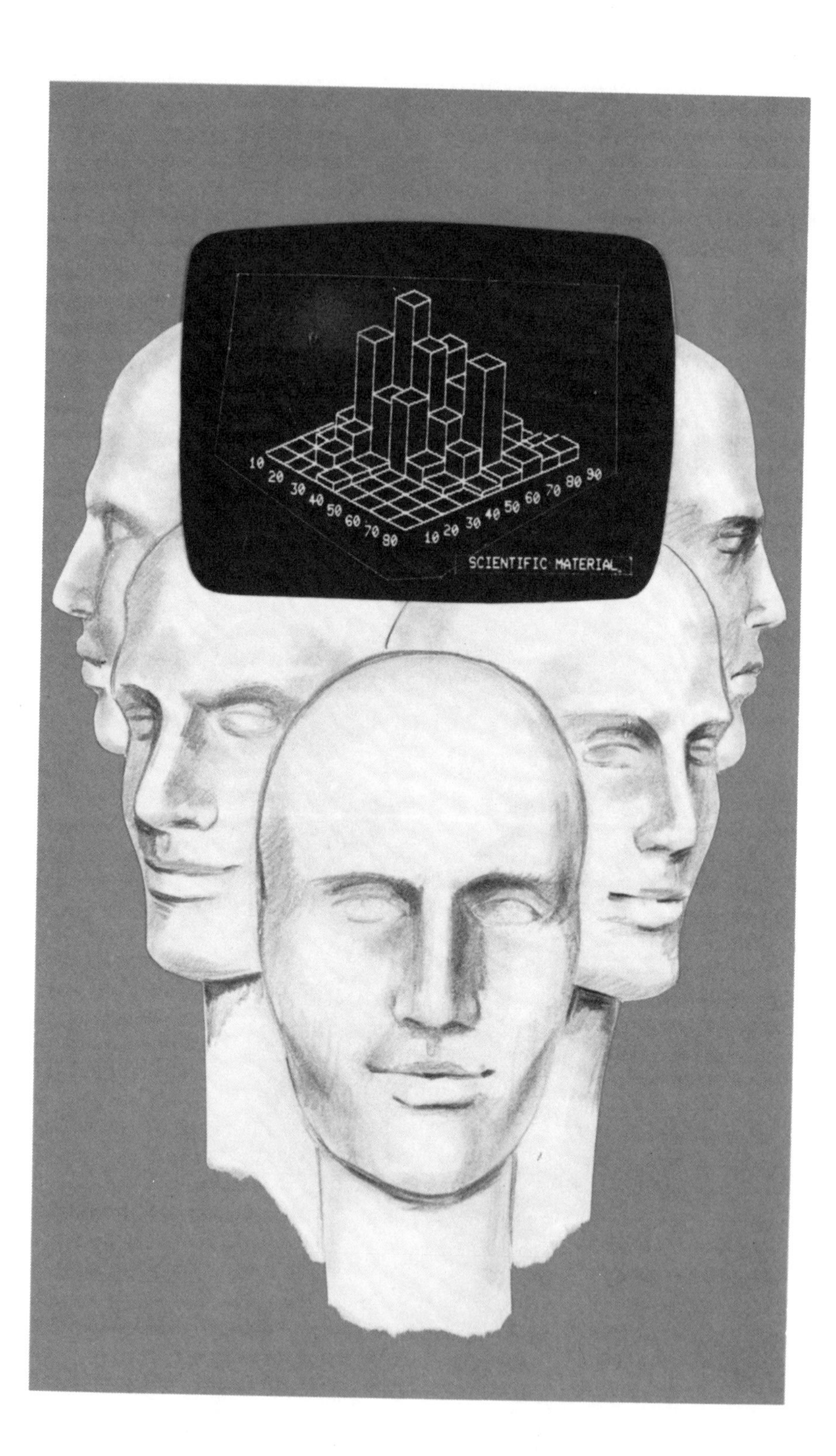
10
20
30
40
50
60
70
80
10
20
30
40
50
60
70
80
90
SCIENTIFIC MATERIAL

6.

IMPACTS ON THE SCIENTIFIC RESEARCH COMMUNITIES

How did the use of EIES for approximately 18 months affect the scientific research communities? In terms of the intellectual and social structure of the group and its ties to other research communities on and off EIES, what happened to communication patterns, cohesiveness, and perceptions of competition in the field? And, most importantly, did EIES in fact help to clarify or resolve theoretical and methodological controversies in the various specialties, as was hypothesized? These questions are the focus of this chapter. Impacts that might be generalizable to any user group, not just scientific research communities, will be the subject of the next chapter. Such more "general" impacts are changes in amount and type of communication, effects on productivity, and general effects on the way that users work and think.

METHODOLOGICAL PROBLEMS AND PROCEDURES

At the end of 18–24 months on EIES, there had been considerable turnover in the composition of many of the scientific groups, with drop-outs and inactives replaced by new members, a portion of whom in turn were inactive and replaced. Thus, even though the size and discipline of a group were the same at post-use as at pre-use, it was a different group because the individuals belonging to it had changed. This is true for a longitudinal study of any scientific community because healthy research communities have new members join and older members retire or stretch their energies into new specialties. However, the rate of replacement was exceptionally high for the online communities.

There are two approaches to the data analysis, "cross-sectional vs. "longitudinal" or "panel." The cross-sectional approach compares the distribution of all responses to the same questionnaire item at two points in time. This gives us a large enough number of cases to permit examination of changes broken down by group. The problem is that we cannot know to what extent differences were produced because the scientific community changed, or because a somewhat different set of individuals responded.

The "panel" approach examines the responses only for those individuals who answered the full set of questionnaires. This reduces the number of cases to a small number for most analyses, unfortunately, resulting in no statistical significance. However, we can see if individuals actually changed over the period of the teleconference.

We shall generally rely on the cross-sectional data describing the research communities at two points in time, since this does not require us to eliminate so much data that the cases remaining are insufficient in number to reach any conclusions. The panel data will be discussed, however, in order to supplement the cross-sectional data with the available information about how specific respondents changed their perceptions of their research specialties over time.

CHANGES IN COMMUNICATION AND COHESION WITHIN THE SCIENTIFIC COMMUNITIES

The majority of EIES users report spending more time communicating with other members of their scientific community as a result of use of the system (Table 6.1). As would be expected, this is strongly related to the amount of time they spend online. At the lowest level of system use half spend less time communicating with their group than they did before it was available. Among the heaviest EIES users, 90% have invested more time in communications. Perhaps the most surprising aspect of the reported changes, however, is how little time some scientists spend communicating within their specialty community. For more than a quarter of the scientists who spent less than 20 hours online over an eighteen to twenty-four month period, this is reported to be more time than they would otherwise have devoted to communication with their peers.

BROADENING OF CONTACTS RATHER THAN ENCAPSULATION

One question asked at the beginning of this research project was whether the use of EIES might not "encapsulate" the communications of its members within the relatively tiny online group of peers. Such a process

TABLE 6 · 1 ***Relative Time Investment in Communication with Specialty Group, by Hours Online***

Hours	Less	More	Same	N
1–19	50%	27%	23%	26
20–49	43%	43%	13%	30
50–99	33%	63%	4%	24
100+	10%	90%	0	19
All	36%	53%	11%	99

Source: Post-Use Questionnaire
Chi-square = 21, p = .01
Contingency coefficient = .42
Question: Compared to the conventional means of communicating with your group, has EIES:
Involved less of your time
Involved more of your time
Involved the same amount of time

would have the negative effect of gradually decreasing contacts with researchers in other specialties and thus impede the valuable and fortuitous process of cross-fertilization of ideas.

On the contrary, EIES is more likely to broaden contacts with local colleagues, as system members become indirect links between the online and off-line worlds. Table 6.2 shows that, for three quarters of the scientists, access to EIES has no effect on the amount of communication with other scientists in the specialty who do not have system access. There are practically no reported instances of a perceived decrease in communications with non-EIES colleagues as a result of system use. However, a significant minority, surprisingly even among those who do not spend much time online, report that communication with these colleagues has actually increased. The explanation is probably that they are serving to relay information about and from the system to off-line colleagues.

Scientists using EIES are much more likely to report an increase in "communications with researchers in other disciplines or specialty areas" rather than a decrease (Table 6.3). There are no statistically significant differences among groups for this finding, though the percentage reporting an increase did vary from only 17% for Group 54 to 54% for Group 40. There is a moderately strong relationship with time online,

TABLE 6 · 2 ***Impact on Communication with Colleagues in the Specialty but Not on EIES, by Hours Online***

Hours	Increased	Decreased	No change	N
1–19	36%	0	64%	28
20–49	24%	3%	73%	33
50–99	8%	4%	88%	25
100+	32%	5%	63%	19
All	25%	3%	72%	105

Source: Post-Use Questionnaire
Chi-square = 7.2, p = .30
Contingency Coefficient = .25
Question: Has the use of EIES affected your communication with any of the following?
Colleagues in your specialty but not on EIES
(Checklist – Increased, Decreased, No Change)

TABLE 6 · 3 ***Increase in Communications with Researchers in Other Disciplines or Specialty Areas***

Cumulative Hours	Increased	Decreased	No Change	N Responding
1–19	30%	0	70	27
20–49	38%	3	59	32
50–99	48%	0	52	25
100+	68%	0	32	19
All	44%	1	55	103

Source: Post-Use Questionnaire
Chi-Square = 9.7, p = .14
Contingency Coefficient = .29

as would be expected. Given our data on the large amount of electronic migration among groups and conferences that took place, most of this perceived increase in communication across disciplines is probably online rather than off.

CHANGES IN PERCEPTION OF AN INTELLECTUAL MAINSTREAM

For all except Group 45 (Devices), the proportion of scientists believing that an "intellectual mainstream" exists in their specialty increased (Table 6.4). Whatever role EIES may have played in clarifying the theoretical and methodological controversies in the fields, it apparently led some group members to feel that they were a little closer to a dominant "paradigm" in their fairly new and interdisciplinary areas. The changes were not very large, however, and the reversal in Group 45 shows that this is a contingent sort of development. As will be described in more detail below, Group 45 is the one in which there was also an increase in perceived competitiveness and in which there were very few perceived advances in clarifying theoretical and methodological issues.

Turning to the panel data, we can compare individual answers to the question of whether the scientists felt more "in" this mainstream or more isolated as the EIES trials progressed. This was measured on an ordinal scale on the pre-use and follow up questionnaires. The scale was:

1 = Completely in the mainstream
2 = Somewhat in the mainstream
3 = Neither in the mainstream nor isolated
4 = Somewhat isolated

TABLE 6 · 4 ***Changes in Perception of an Intellectual Mainstream, by Specialty Group (cross sectional data)***

	Pre Use		Post Use	
Group	% yes	N	% yes	N
30(Futures)	54%	13	63%	19
35(Soc Nets)	27%	22	31%	26
40(Gen Sys)	33%	30	41%	29
45(Devices)	71%	14	56%	18

Note: Group 54 has too few post-use responses for comparison.
Question: Is there a commonly accepted "intellectual mainstream" in the specialty?

5 = Completely isolated

There were only 53 cases with both pieces of data. Contrary to expectations, there is a significant shift toward feeling *less* in the "intellectual mainstream" of the specialty. The mean at time one was 1.3, and at time two, 2.3 ($T=4.78$, $p=.01$). This is a rather surprising finding, and one can only speculate on the reasons. Perhaps the online subgroup recognized its discussions and conclusions as separating them from accepted or taken for granted theories or priorities shared by the rest of the "off-line" world in their specialty.

PERCEPTIONS OF SPECIALTY GROUP COHESIVENESS

The follow-up and post-use questionnaires asked whether participants would describe their groups as

- more a collection of individuals than a research community;
- a set of cliques or subgroups with interests and activities in common, but not an integrated community; or
- a well-integrated research community that shares many interests and activities in common.

Some groups became more integrated and some less cohesive. Overall, there was no significant shift in perception during the course of the operational trials. At both times, only a minority perceived their groups as integrated research communities. This lack of increased cohesion as subjectively perceived fits in well with the behavioral data that showed that many group members "migrated" to more active or compatible groups on line (See Chapter 2).

CHANGES IN PERCEPTIONS OF COMPETITION

However, the panel data indicate that there is a selection process at work as well as a change in attitudes among individual members. Before and after using EIES, participants were asked, "How would you rate the degree or intensity of competition within your specialty?" (Very intense, intense, moderate, low or non-existent were the response choices provided). About the same proportion of most groups reported very intense or intense competition at both points in time. The exception is Group 45 (Devices). It started out with only 11% perceiving very intense or intense competition and ended up with 43%.

The consistent shifts which do occur are concentrated within specific kinds of competition. There is a dramatic increase in all groups in perceived scarcity of or competition for funds (for instance, from 18% to

47% for the Futures group; from 21% to 82% for the Devices group). Those who perceived unethical behavior among their peers dropped out of EIES and did not respond to the post-use questionnaire, so this reason practically disappears. There is also some increase in perceptions of strongly opposing views.

The panel data on the 45 to 53 individuals who answered both questions reinforce the results of the cross-sectional data. On the question on overall degree or intensity of competition, the mean was 3.2 on the one to five scale used, at both points in time. ($T=.17$, $p=.8$). For specific kinds of competition, there was a significant increase for the same types that are apparent in the cross sectional data: competition over funds, perception of rival groups, and strong opposing views. (For example, with "yes" coded as "1" and no check of a reason coded as 2, the mean for "opposing views" was 1.7 at pre-use and 1.4 at post-use; $T=3.5$, $p=<.05$). Thus, the implication derived from the cross-sectional data that increases in perceptions of specific kinds of competition within the specialties were caused by the discussions and interactions on the system is supported by the panel data.

BETTER UNDERSTANDING OF OTHERS' WORK

The majority of EIES users agree that the increased communication with peers has changed their understanding of the interests and activities of other scientists in the specialty. The more time they spend online, the more likely it is that such increased understanding will occur (Table 6.5). There are significant differences among the specialty groups in the extent to which this process occurs. Such impacts are not related to our other measures of group success: the two groups in which there is the most "increased understanding" reported include one of the most successful (General Systems) and one of the least successful (Mental Workload).

About half of the scientists report the related perception that use of the system has changed their views of how their own work relates to that of others in the specialty. Those who spend at least 50 hours online are most likely to report this perception.

CLARIFICATION OF THEORETICAL AND METHODOLOGICAL CONTROVERSIES

One fundamental question about technology such as EIES is whether it can speed the development of a disciplinary paradigm or the process of paradigm change. Especially intriguing are specialties where a previous theoretical and methodological framework that has been dominating the

TABLE 6 · 5 ***Extent to Which EIES has Changed Understanding of Others in Specialty, by Hours Online***

Hours	Strongly Agree	Agree	Neither	Disagree or Strongly Disagree	#
1–19	3%	45%	17%	35%	29
20–49	15%	42%	36%	6%	33
50–99	16%	28%	22%	7%	25
100+	21%	53%	21%	5%	19
All	13%	48%	26%	13%	106

Source: Post-Use Questionnaire
Chi-square = 20.1, p = .02
gamma = .34
Question: EIES has changed my understanding of the interests and/or activities of others in my specialty.

field seems inadequate for answering fundamental questions or guiding fruitful research. There are at least three parts of this process that can be identified: formation of new approaches, clarification of the differences between the old and the new approaches, and resolution of the controversy by synthesis or replacement, or through the demise of the proposed new approach.

Overall, about half of EIES users felt that the use of the system had clarified theoretical controversies within the field. It was generally not felt that there had been a "great deal" of clarification, but only that there had been "some." Many of the comments accompanying this section of the post-use questionnaire pointed out that the controversies among competing theoretical positions had been clarified, but not resolved. The amount of progress on theoretical conflicts varied by specialty, with General Systems Theory (Group 40) reporting the most progress, and Devices for the Disabled, a relatively applied and non-theoretical discipline, the least (see Table 6.6). As would be expected, perception of clarification of theoretical controversies is very strongly related to amount of time spent online. Almost all of the heavy users of the system felt that this was one outcome of their use of EIES, whereas those who had spent less than an hour a month online on the average obtained no such benefit.

Table 6.7 lists some of the specific theoretical issues which were

TABLE 6 · 6 ***Clarification of Theoretical Controversies, by Specialty Group***

Group	Great Deal	Somewhat	No	#
30 (Futures)	6%	50	44	18
35 (Social Networks)	4%	44	52	25
40 (General Systems)	14%	52	34	29
45 (Devices)	5%	16	79	19
54 (Mental Workload)	0	57	43	7
All	7%	43	50	98

Source: Post-Use Questionnaire
Chi-square = 11.9, p = .15
Question: Has EIES helped to clarify any theoretical controversies in the specialty area?
–yes, a great deal
–yes, somewhat
–no

named by participants in the various specialty groups as having been clarified. Group 30 (Futures Research) could not come up with very much specific. Group 40 (General Systems Theory), which generated the largest percentage overall of perceived progress on theoretical issues, focussed mainly on the open vs. closed system paradigms.

Generally, use of EIES was seen as somewhat less likely to have helped clarify methodological controversies than theoretical issues in a scientific specialty. However, this was not true in all groups. Social network theory members were more likely to perceive methodological progress than theoretical progress, and named several specific methodological clarifications. (See Table 6.8.)

SUMMARY AND CONCLUSIONS

As a result of using EIES for a period of 18–24 months,

A. Total communication within the scientific specialty increases.

1. The majority of EIES members spend more time communicating with their specialty group colleagues than they otherwise would.

TABLE 6 · 7 ***Nature of Perceived Theoretical Clarifications, by Group (Representative Paraphrases)***

Group 30 (Futures Research)

1. Cross impact (expect a paper to be written)
2. Exploring concepts of decentralization
3. Subjective probability

Group 35 (Social Network Analysis)

1. Clarification of differences in approaches to structure.
2. On the issues of cognitive salience of networks, the conference has helped by expanding the controversy (getting opposing views out in the open).
3. We have a clearer idea of the areas in which there is diversity of conceptualization and interpretation and where more work needs to be done.
4. Concept of centrality and its measurement has been clarified, but not resolved.

Group 40 (General Systems Theory)

The open system/closed system debate helped to clarify the difference between general systems theory and cybernetics.

It has sharpened the issues involved.

Group 45 (Devices for the Disabled)

1. Problems of marketing and commercializing devices—not resolved at all.
2. The kind of information that needs to be sent to government policy makers.

Group 54 (Mental Workload)

1. Information theory measures. Man-machine design. (not resolved)
2. Definitions/limitations

TABLE 6 · 8 ***Clarification of Methodological Controversies, by Specialty Group***

Group	Great Deal	Somewhat	No	# (100%)
30(Futures)	0	50	50	18
35(Social Nets)	8%	46	46	26
40(Gen Sys)	0	31	69	29
45(Devices)	5%	21	74	19
54(Mental Workload)	0	50	50	6
All	3%	38%	59%	98

Source: Post-Use Questionnaire
Chi-square = 9.1, p = .32
Question: Has EIES helped to clarify any methodological controversies in the specialty area?
–yes, a great deal
–yes, somewhat
–no

2. Three-quarters report no change in amount of communication with off-line colleagues in the specialty. One-quarter report an increase. Thus, there is an expansion of indirect communication ties, rather than an "encapsulation" of the online group. This lack of negative impact on communication with off-line colleagues is an important finding.
3. Almost half report an increase in communication with scientists in other specialties or disciplines, and practically none report a decrease.

B. As a result of this increase in communication:

1. There is not an increase in the perceived degree of integration within the specialties. At the end of the observed period of EIES use, as at the beginning, the specialties are generally seen as "collections of individuals" or "sets of cliques," rather than as well integrated research communities.
2. For some groups there is an increase in the extent to which the scientists perceive an "intellectual mainstream" or commonly recognized paradigm in the specialty.
3. There is a significant change in the extent to which the individual

scientists perceive themselves as "in" such a mainstream, to the extent that one exists. The change is toward perceiving themselves as farther "out" of the mainstream. This is an unexpected result.

4. There is a tendency for perception of increased competitiveness within the groups related to rivalries, conflicting ideas, and limited funds.
5. There is an increased understanding of the interests and activities of other scientists in the specialty, and of how one's work relates to theirs.

About half the scientists feel that use of the system has somewhat clarified theoretical controversies within their specialties. Such perceptions vary significantly among the specialty groups and increase with time online. Clarification of methodological controversies is less frequently perceived. Resolution of the controversies has not occurred.

Perhaps the increased communication that occurs on EIES has effects like those of a political campaign. One becomes more aware of the issues on which there is disagreement, and of the divisions within the (scientific) society. And perhaps one needs a structured process like an "election" to resolve these disagreements.

The findings reflect the participants' judgments that EIES is better for generating ideas and exchanging opinions than for resolving disagreements. However, controlled experiments indicate that it is possible to create structured processes of communication within the medium that do make it likely that a group will resolve its differences and reach consensus. Either formal human leadership processes, or a decision aid based on systematic computer feedback on the nature of differences of opinion as expressed through formal "voting," have enabled groups on EIES to reach total consensus (See Hiltz, Johnson, and Turoff, forthcoming). It would be interesting to see if future groups of scientists could resolve the controversies which surface as a result of their computerized communication with the assistance of such special structures for generating consensus.

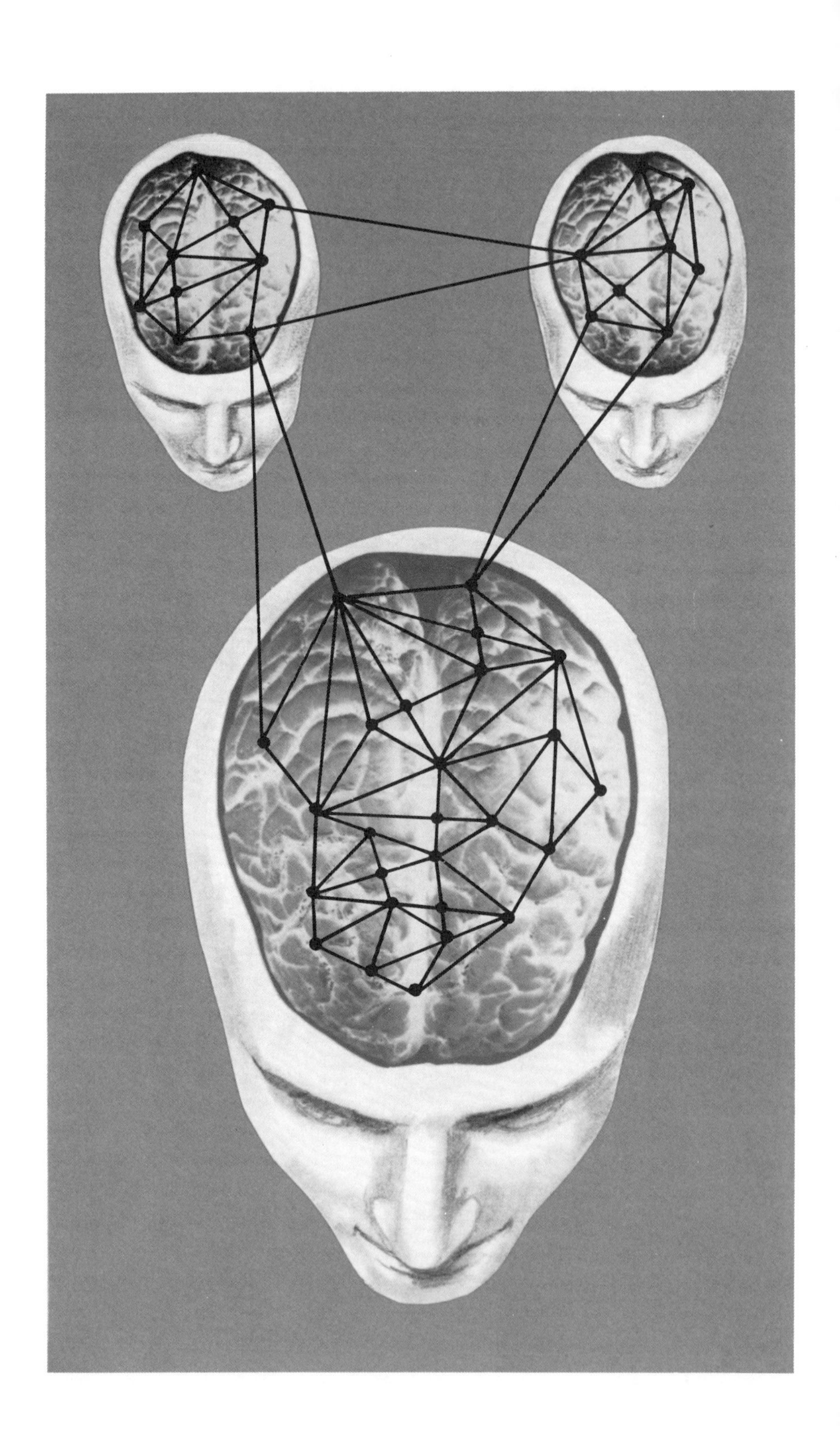

7.

IMPACTS ON COMMUNICATION AND PRODUCTIVITY

The major justification for computer-mediated communication systems is increases in the productivity of the "knowledge workers," who now make up the majority of the labor force in the United States and other "post-industrial" societies. For example, James Olsen, vice chairman of AT&T, states that "the merger of two great technologies—computers and communications—will present tremendous opportunities for the resumption of productivity advances" in America (quoted in Oreskes, 1981:6). Some savings may be obtained by replacing telephone calls, typed letters, memos, and even some meetings with communication via this generally cheaper, faster medium. One estimate (by Bair, in Uhlig, Farber, and Bair, 1979:361) is that labor savings for managerial, professional, and technical workers making optimal use of such systems could equate to two hours a day per person, or $62 billion dollars per year based on total current U.S. annual labor costs in these areas.

Even more significant than these cost-displacement savings obtained through greater efficiency, it is argued, are the potential increases in productivity due to more effective working patterns. The reports and decisions produced by knowledge workers may actually be significantly better. For example, Johansen, DeGrasse, and Wilson (1978:1) report that computer conferencing facilitated the work of teams of geologists by reducing delays in information exchange, coordinating the roles of remotely-located participants, and "generally improving the ability to deal with large amounts of information."

We will look first at three observed changes in the work patterns of the scientists. A section on media substitution shows how EIES use affects the use of other communication modes, such as mail and telephone. Substitution is of interest because it provides opportunities for cost savings by displacing the use of less efficient media. The second process examined is increases in "connectivity" (the size and density of communication networks) reported by those who made substantial use of the system. Then we will turn to scientists' descriptions of how the system changed the way in which they think and work. The fourth section looks at changes in activities related to productivity reported by the participants after using the system for approximately two years. Finally, a multivariate analysis traces the causal relationships among productivity-related processes and outcomes.

Previous studies of the effects of PLANET use on energy researchers (Johansen, DeGrasse, and Wilson, 1978) and of impacts of the Swedish COM system on its users in five geographically dispersed offices of the Swedish National Defense Research Institute (Palme, 1981) will be used to make preliminary estimates of the generalizability of the findings to other computerized conferencing systems. Subsequent studies on other users and other systems will be necessary to determine the overall generalizability of the findings.

MEDIA SUBSTITUTION?

One possible expectation is that a computerized conferencing system can SUBSTITUTE for communication via other media, taking the place of mail, telephone, or face-to-face meetings. In the case of scientific communities, information exchanges online might conceivably even substitute for book or journal reading, in the sense that the time invested in reading papers and conferences online might be subtracted from some fixed total amount of time available for "keeping up" with the professional literature in one's field. Under the substitution model, one would expect a decrease in the use of other media.

Some of the greatest hopes for economic viability of computer-based communication systems stem from the idea that it may replace more expensive means of communication. Nilles et al. (1976) focus on the ability to telecommute to work rather than waste time and petroleum resources on daily commutation to an office. Kollen's (1975) study looks at "travel/communication tradeoffs" mainly in terms of substitution for business trips at which face-to-face meetings take place.

One of the stated objectives for the use of message systems is usually to replace the letter or the internal memo or the telephone call. For instance, one goal/justification of the electronic mail system tried by Owens Corning, as reported by McNurlin (1980:2–3) was cost displacement through substitution for mail and telephone:

> . . . experienced users typically replaced four to six communications a day, which, with a future projected population in the company of 1500 users, would show replacement savings of $600,000 a year.

On the other hand, perhaps computer-mediated communications are ADDED ON to other communications rather than substituting a new mode. This may be particularly true if a system includes only a relatively small number of addresses or members, so that most of the people with whom one communicates are not available online. Under these circumstances the user might maintain usual communications channels but add on to them new communication with people who have not previously been easily accessible. Under the add-on model one would expect to see use of other communication modes remain constant ("no effect").

A third hypothesis might be termed communication EXPANSION. This model pictures CC being added on to existing communications; and then stimulating more communications via other media. This might take the form of telephone or travel or mails to supplement CC comunication with people met on EIES, increased reading of books or journals due to discussions and references encountered online, or increased communication with off-line colleagues that is stimulated by system use. Under the expansion model, one would expect to see that use of other media actually increases.

Whether substitution, add-on, or expansion phenomena are observed will vary with amount of system use. Low levels of use should not affect other communications modes very much. It is probably the EIES users who spent a relatively high amount of time online (100 hours or more over eighteen to twenty-four months) who are most predictive of the potential media substitution effects, should such systems become widely used within an organization or interest community. Thus, we will look at reported effects cross-tabulated by amount of time online. To the extent that significant differences are observed among the user groups on EIES, it indicates that media substitution effects are also dependent on application. (Task, size, and social cohesion of the group, etc, are all bound up in differences among the groups on the EIES system).

MAIL AND TELEPHONE

In Tables 7.1 and 7.2, we see that there is generally an "add-on" effect in relation to mail and telephone, but as system use increases, the "substitution" effect becomes more prominent. Overall, a quarter of all members and half of the heavy users report a decrease in telephone use, as a result of EIES. However, a minority demonstrate an "expansion" effect: 14% overall report an increase in telephone use attributable to using

TABLE 7 · 1 ***Impact on Amount of Use of Telephone, by Hours Online***

Hours	Increased	No effect	Decreased	N
1–19	11%	71%	18%	28
20–49	6%	81%	13%	32
50–99	24%	52%	24%	25
100+	17%	33%	50%	18
All	14%	63%	23%	103

Source: Post-Use Questionnaire
Chi-square = 16, p = .01
Gamma = .14
Question: Has the use of EIES changed the amount of your use of other media in the last year? (Media checklist with increased—No effect—Decreased as choices)

TABLE 7 · 2 ***Impact on Amount of Use of Mail, by Hours Online***

Hours	Increased	No effect	Decreased	N
1–19	11%	68%	21%	28
20–49	19%	47%	34%	32
50–99	32%	28%	40%	25
100+	22%	28%	50%	18
All	20%	45%	35%	103

Source: Post-Use Questionnaire
Chi-square = 11.9, p = .06
Question: Has the use of EIES changed the amount of your use of other media in the last year? (Media checklist with increased—No effect—Decreased as choices)

EIES, and this increase is also directly related to amount of system use.

The pattern for mail is similar, but stronger. At low levels of system use, there is most likely to be "no change" in the use of mail. But the likelihood of both reported decreases (substitution of computerized communications for mail) and of reported increases (more mail as a result of system use) varies directly as a function of the amount of system use. Among medium to heavy users, decreased mail is the modal pattern; but expanded mail use also occurs more frequently.

A probable explanation is that online communication substitutes for some mail or telephone but stimulates other contacts that might not otherwise take place. For instance, users may apprise one other of available preprints or other documents, which are then sent by mail. If a subject of mutual interest is likely to take a great deal of discussion, participants who find themselves online at the same time frequently decide to talk it over on the telephone to resolve an issue or to get another set of cues about each other's feelings. In other words, qualitative observations suggest that dyads resort to the telephone as a supplementary means of communication for fairly long (ten minute or more) conversations, particularly if they find each other online at the same time and are thus obviously available to take a call. It is the heaviest users and those who make the most new contacts who are most likely to expand their use of mail and telephone as a result of computerized communications.

Analysis of the patterns for mail and telephone by group indicate that the prevalence of substitution, expansion, or add-on effects are dependent on the group context. Among EIES users, nobody in Group 30 (Futures) reported an increase in the use of mail or telephone as a result of using EIES. This futures research group had the largest, most active conference, and thus a great deal of group rather than dyadic communication, for which mail and telephone are most suitable. By contrast, the largest proportion reporting an increase in the use of the telephone as a result of using the system occurs for the Mental Workload group, which had the least successful conference.

The findings of the PLANET study are comparable in the sense that they observe that "there is no simple relationship between computer conferencing and use of conventional media" (Johansen, DeGrasse, and Wilson, 1978:69). In most cases, however, the frequency of the reported use of mail decreased during the period of observation. On the other hand, telephone use was as likely to increase as to decrease. In both the EIES and PLANET studies, there were some differences in observed patterns among user groups. Looking at the two sets of results, one can project that substantial use of such systems is likely to result in substitution

TABLE 7 · 3 ***Impact on Amount of Travel to Professional Meetings, by Hours Online***

Hours	Increased	No effect	Decreased	N
1–19	7%	83%	10%	29
20–49	7%	81%	13%	31
50–99	12%	88%	0	25
100+	17%	61%	22%	18
All	10%	80%	11%	103

Source: Post-Use Questionnaire
Chi-square = 7.7, p = .26
Contingency Coefficient = .26
Question: Has the use of EIES changed the amount of your use of other media in the last year? (Media checklist with increased—No effect—Decreased as choices)

for the mails, but that the substitution of computerized communication for some telephone calls is likely to be balanced by the stimulation of other telephone communication as a result of new contacts and/or unresolved issues that emerge through communication on the system. In any case, there is likely to be considerable variability among user groups, so that one cannot make accurate predictions for a specific group.

TRAVEL/TELECOMMUNICATIONS TRADEOFFS?

Turning to travel substitution, attendance at professional meetings was separated from personal visits with distant colleagues. System use does not significantly impact attendance at professional society meetings. As shown in Table 7.3, 80 percent report "no effect," and those who do perceive an effect are almost as likely to report an increase as a decrease, at all levels of system use. In terms of travel for a personal visit (Table 7.4), there is more likely to be a perceived impact, and once again, such travel is about as likely to increase as to decrease. Among the heaviest users of the system, almost a quarter report an increase in travel for this purpose. Apparently, as long as travel budgets are not cut, contact with colleagues online is about as likely to stimulate travel as to substitute for it. Anecdotal evidence suggests that among those who interact a great deal online but have never met in person, curiosity prompts extensions to business or personal trips made for other purposes, in order to meet with one's online acquaintances.

TABLE 7 · 4 ***Impact on Visits with Researchers in Other Locations, By Hours Online***

Hours	Increased	No effect	Decreased	N
1–19	11%	82%	7%	28
20–49	13%	69%	19%	32
50–99	8%	88%	4%	25
100+	22%	50%	28%	18
All	13%	74%	14%	103

Source: Post-Use Questionnaire
Chi-square = 10.1, p = .12
Contingency Coefficient = .30
Question: Has the use of EIES changed the amount of your use of other media in the last year? (Media checklist with increased—No effect—Decreased as choices)

The PLANET study also showed mixed results on travel substitution. Two of their groups showed some increase in travel for communication with other researchers over the period of the study, and two showed a decrease (Johansen, DeGrasse, and Wilson, 1978:74–75). They also conclude that CC has the potential to substitute for travel to face-to-face conferences if researchers feel overburdened by "too much" travel or have insufficient travel funds; ; but that the new communications channels opened up by such a system when it is on an international network may also actually stimulate travel for meetings that otherwise would not have taken place.

READING

Reading professional books and journals is much more likely to increase rather than decrease as a result of using EIES (Table 7.5). apparently the discussions with one's colleagues lead to more interest in reading journals, since the more time spent online, the more likely it is that reading increases.

However, this effect was found to be group dependent. The proportions reporting an increase in reading varied from only 6% in the Devices for the Handicapped Group, to 40% in the General Systems Theory group. And for the PLANET study, two of the four groups showed a definite decrease in the reported frequency of reading work-related books and articles (Johansen, DeGrasse, and Wilson, 1978:76–77).

TABLE 7 · 5 ***Impact on Reading Journals or Books, by Hours Online***

Hours	Increased	No effect	Decreased	N
1–19	17%	75%	7%	29
20–49	25%	63%	13%	32
50–99	32%	64%	4%	25
100+	44%	39%	17%	18
All	28%	62%	10%	104

Source: Post-Use Questionnaire
Chi-square = 7.9, p = .24
Contingency Coefficient = .27
Question: Has the use of EIES changed the amount of your use of other media in the last year? (Media checklist with increased—No effect—Decreased as choices)

TABLE 7 · 6 ***Impact on Communication with Colleagues in One's Own Organization, by Hours Online***

Hours	Increased	Decreased	No Change	N
1–19	43%	0	58%	28
20–49	15%	6%	79%	33
50–99	20%	8%	72%	25
100+	16%	0	84%	19
All	24%	4%	72%	105

Source: Post-Use Questionnaire
Chi-square = 10.8, p = .09
Contingency Coefficient = .30
Question: Has the use of EIES effected your communication with any of the following? Colleagues at your institution or organization.

COMMUNICATION WITH CO-LOCATED COLLEAGUES

Use of EIES is more likely to increase than decrease communication with one's co-located (off-line) colleagues (Table 7.6). It is surprising that even the lowest level users are likely to report an increase in communication with colleagues within their own organization as a result of using EIES. Practically no one reports a decrease in communication with co-located colleagues as a result of using the system. Perhaps the large proportion of low level users who report an increase in local communication can be explained by their use of the system as a kind of toy which they occasionally demonstrated to colleagues as a curiosity or status symbol. Since we did not ask about the content of off-line communications that were increasing for any of the modes, however, we can only speculate about the nature of it. It is likely that users become a link for their colleagues between the online and off-line communications networks.

SUMMARY

In sum, for all modes of communication, low levels of system use are most likely to have no effect on the use of other communication media; system use is simply added onto existing communication. However, at high levels of system use, one is very likely to also expand the use of other communications media as an adjunct to online communications. This corresponds fairly well to the previous findings of a study of PLANET users.

The findings of the COM study also indicate that media substitution is not the usual effect of use of this medium, but rather that it stimulates much "new" communication which would not otherwise take place. An inquiry to a random sample of writers of COM messages showed that 7% replaced letters or memos, 23% replaced telephone calls, 17% replaced face-to-face meetings, 3% replaced other media, but 50% were "new communication which would not have occurred without COM" (Palme, 1981:22).

There is very little tendency for system use to simply substitute for other media. On the contrary, all indications are that using such a system will significantly increase the total volume of communication. The content and consequences of this increase in communication are the focus of the remainder of the chapter.

INCREASED CONNECTIVITY

There are many indications that EIES use expands the size and density of social networks. By size, we mean the total number of persons with whom one is directly or indirectly in contact, and with whom one can fairly easily exchange information, ideas, or more personal communications. By density, we mean the number of connections within the social network. Density is defined mathematically as the actual number of ties among pairs in a network, divided by the total possible number of ties between pairs. So, for instance, a density of .50 would mean that half of the pairs in the social network or group are connected. Another concept is intensity or multistrandedness of relationships. There are many kinds of ties, from knowledge or awareness of one another to close personal friendship. It is hypothesized that systems such as EIES can increase the intensity or strength of ties as well as the size and density of networks. Such large, densely knit networks with many rich (multistranded) relationships are potentially a very fruitful social setting for scientific progress or other kinds of "knowledge work."

The questionnaire data measuring growth in social networks (Table 7.7) shows that most EIES users have met and gotten to know other scientists over EIES. As would be expected, the number of new social ties established on EIES is highly correlated with the amount of time spent online. Among those who had spent 100 hours or more online, a third had expanded their social/scientific network by eleven or more new persons.

THE GROUP 35 SOCIAL NETWORKS STUDY

Twenty-nine members of this group completed an online version of a social networks questionnaire at the start of the experimental period, and a mailed version seven months later. Since twenty-one of the twenty-

TABLE 7 · 7 ***Number of Persons Met on EIES, by Hours Online***

Hours	None	1–5	6–10	11 or more	N
1–19	52%	35%	10%	3%	29
20–49	27%	37%	17%	20%	30
50–99	20%	48%	20%	12%	25
100+	6%	50%	11%	33%	18
All	28%	41%	15%	16%	102

Source: Post-Use Questionnaire
Chi-square = 23, p = .03
Gamma = .38
Question: "How many different people do you feel you are actually exchanging information or communicating with on this system, currently?
Of these, how many have you "met" (gotten to know) over EIES?"

nine had attended a two-day face-to-face meeting just before they completed the first questionnaire, it is surprising that a little less than half reported ever having "met" one another.

There were four types or levels of intensity of relationship asked for at the two points in time. Each participant was asked to designate those they had heard of or read publications by; those they had met, or exchanged letters or phone calls or computer conferenced with; those whom they considered "friends;" and finally, those whom they considered "close personal friends." Table 7.8 shows the density for these four

TABLE 7 · 8 ***Increases in Connectivity as a Result of Using EIES***

A. Density of Four Types of Social Relations Before and After Seven Months of Using EIES, Group 35

	TIME	
Relation	First	Second
Heard of	.62	.77
Met	.49	.68
Friends	.14	.22
Close friends	.05	.06

Source: Freeman and Freeman, 1980:79

B. Average distances between reachable pairs and number of reachable pairs for four relations at two times, Group 35

	TIME			
	FIRST		SECOND	
Relation	Distance	No. of Prs.	Distance	No. of Prs.
Heard of	1.38	812	1.17	812
Met	1.52	812	1.30	812
Friends	2.76	728	2.18	812
Close friends	2.01	96	3.13	221

Source: Freeman and Freeman, 1980:81

levels of relationships at the two points in time. The density coefficients correspond to the proportion of the group members responding who have the direct tie; for example, it shows that 14% considered themselves to be "friends" before the online experience began, and 22% considered themselves "friends" seven months later:

> There were noticeable increases in the proportion of people reporting relationships of all four kinds. It would seem that the computer conference, or perhaps some other events that took place during that seven month period, brought these people closer together (Freeman and Freeman, 1980, p. 80)

The analysis also measures distance or "reachability" among Group 35 members. A person is reachable if one is linked directly or indirectly through several ties (such as friend's friend). Distance is the number of links required to reach someone by the shortest route. For example, my friend is one link away, my friend's friend is two links (or d=2) away. They found that the number of reachable pairs grew whenever possible (when it had not already reached 100% for the 812 possible pairs), and that the distances were shrinking on all relationships except those of close personal friends (Table 7.8).

The entire group was connected by a friendship network by the end of seven months; almost everyone who was not a friend was a "friend of a friend." (The distance of just over two links on the average indicates this). Such feelings of friendship can be very valuable when asking for favors, such as a review of a draft article or information about a topic of interest. The Freemans conclude that the network was changing from a clique structure to a genuine community (Freeman and Freeman, 1980:80):

> For close personal friends, data from the first questionnaire seemed to show the presence of tight little cliques; by the time of the administration of the second questionnaire there were many more personal friends reported and they were beginning to be loosely linked together into larger structures. This suggests that at the end of the second questionnaire there was much more of a "community" among these social networks people.

In sum, detailed "before and after" social network data for one of the scientific communities using the system confirm that substantial increases in connectivity can take place as a result of communication through this new medium.

Other studies also support this finding. For instance, Panko and Panko (1980) report that the strongest experienced benefit reported by the users of the DARCOM electronic mail system was an overall increase in communication with distantly located colleagues. And for the synthesis project on evaluations of computer-mediated communication, all nine of the evaluators with data on this potential impact reported that it supported the finding that the medium causes an overall increase in communication links (Hiltz and Kerr, 1981:212–214).

A detailed study of the COM system in Sweden also supports the finding of increased connectivity. Palme (1981:1) states that

> COM has for many of its users meant an increase in the circle of people with which they exchange experience and ideas on a daily basis, and has meant that information and viewpoints both can be disseminated and collected to and from more people faster than was possible before.

IMPACTS ON THE WAY IN WHICH MEMBERS THINK AND WORK

An open-ended question probed the extent to which the use of EIES had "any impacts on the way in which you think and work, in general." Respondents were asked to check yes or no, and then to "describe these impacts in as much detail as possible."

Overall, 52% report general impacts on working patterns, with many describing them (see the examples in Table 7.9). The reported impacts fall into four broad, partially overlapping categories. One has to do with broadened professional perspectives or activities. The second relates to increases in communication or connectivity. The third refers to a kind of change in perspective about the relation of self and cosmos caused by the communications medium: disappearance of space and time are frequently mentioned aspects of this. And the fourth relates to specific work habits, such as being more organized, working at home more, and increased pace of work. These qualitative, subjective reports seem to be related to an enlargement of the cognitive resources available to the scientists, and to be perceived as largely beneficial.

As would be expected, impacts on the way one works and thinks are more likely the more one uses a system such as EIES. Reported impacts increase steadily from 39% of those with less than twenty hours total experience online to 78% of those with 100 hours or more of online time.

PERCEIVED IMPROVEMENTS IN PRODUCTIVITY

Objective general measures of the productivity of scientists and other professionals are difficult to obtain. For the things being produced, such as reports and decisions, "more" is not necessarily "better"; it is the quality

TABLE 7 · 9 ***Impacts on the Way Users Work and Think (Selected Quotes from an Open-Ended Post Use Question)***

A. Broadened Perspectives

It has broadened my perspectives on my own work and on the environment in which I am working. I have been exposed to ideas which I would not otherwise have encountered and have been able to participate in more wide-ranging discussions than ever before. I will miss the intellectual stimulation, the diversity of ideas, and the immediacy of communication.

Much more opportunity to discuss basic intuitions, perspectives and opinions on what is valuable in this field of research. My own work has broadened a great deal as a result.

I have been exposed to (1) a variety of people in my research area previously unknown to me (2) people in other research areas and their ideas about the world (3) I have been able to ask for help from leading members of my research community about current research problems.

It has made me more aware of the issues which some people in the field consider important; this has included some surprises.

Broader exposure to ideas. More aware of controversy within disciplines. Familiarity with people in field.

The world is larger than I thought—positively in that there are actually knowledgeable people out there and (temporarily) negatively in that there are so many with so many ideas—that (temporarily) coherence and holding onto who I am suffer a bit.

B. Increased Communication and Connectivity

My first reaction now is to get on the system and get in touch with the appropriate person. I have been doing more communicating.

More informal contacts

TABLE 7 · 9 *(cont.)*

B. Increased Communication and Connectivity, *(cont.)*

The instantaneous feedback capability of EIES in producing written material has had an outstanding effect on my work . . .

I can kick ideas around among a larger circle of researchers.

I have become addicted to instant gratification of need to communicate. I communicate more often on both important and trivial matters.

C. Less Space and Time Bound

Feel less time-bound

Being on EIES is like being in another space-time. I feel like I am simultaneously in France and in the States, which has been a longstanding dream of mine.

of what is produced that is crucial. It was decided for the purposes of this study to rely upon subjective reports of the scientists about how the process and outcome of their work changed as a result of using the system, rather than, as Johansen, DeGrasse, and Wilson (1978, p. 16) put it, "becoming bogged down in direct and sometimes problematic output measures." The specific process variables used were selected on the basis of previous studies about factors related to scientific productivity.

Seven separate questions on productivity-related factors were included in the post-use questionnaire. The responses to these seven questions are all highly intercorrelated, and a factor analysis shows that there is only one underlying factor or dimension—in other words, they are all measuring different aspects of the same thing, as was hypothesized when the instrument was designed. This means that the responses can be combined into a single index or variable for analysis. The variable will be called PRODUCTIVITY for short, and is formed by adding together the response scores (from one to five, strongly agree to strongly disagree) on the following items:

QUALITY = "Use of EIES has increased my productivity in terms of the quality of work recently completed or underway."

QUANTITY="Use of EIES has increased my productivity in terms of the quantity of work recently completed or underway."

IDEAS="Use of EIES has increased my 'stock of ideas' that might be used in future work."

RELATE="EIES has changed my view of how my own work relates to that of others in my specialty."

INFO="EIES has provided me leads, references, or other information useful in my work."

FAMILIAR="EIES has increased the familiarity of others with my work."

UNDERSTAND="EIES has changed my understanding of the interests and/or activities of others in my specialty."

Table 7.10 summarizes the responses to the individual questions. One cannot take such reports at face value; the respondents may have been overly generous towards EIES and inclined to see improvements in their own work where more objective third parties would not. On the other hand, the nature of intellectual work is such that only the person doing it is in a position to say whether something has helped or not. The responses are shown for all respondents, and for the heaviest users; all items were related to amount of time online (gamma ranged from .26 to .37).

First of all, we note that the system is somewhat more likely to be associated with increases in perceived quality of work than with quantity of work. The ways in which quality is improved can be implied by reports of specific effects such as increasing the "stock of ideas," providing leads and references, and improving conceptual understanding. The latter refers to shared conceptual space: improved understanding of the nature of work being done by one's peers and increases in their familiarity with one's own work. These effects are reported by about half of all users. The largest percentages of reported productivity-related gains occur for increasing the stock of ideas and for providing leads, references, or other information. This enables us to understand the link between increased connectivity and increased productivity, in terms of the content of the communications that are flowing through the new online links which prove helpful to the scientists in their work.

TABLE 7 · 10 ***Perceived Impacts on Productivity***

Item	Strongly Agree or Agree	Neither	Disagree or Strongly Disagree
QUALITY			
100+	68%	21%	11%
All	37%	28%	35%
QUANTITY			
100+	53%	26%	21%
All	27%	31%	42%
IDEAS			
100+	90%	5%	5%
All	71%	8%	21%
RELATE			
100+	47%	32%	21%
All	46%	25%	29%
INFO			
100+	95%	5%	0
All	79%	9%	12%
FAMILIAR			
100+	84%	16%	0
All	54%	32%	14%
UNDERSTAND			
100+	74%	21%	5%
All	61%	26%	13%

Source = Post-Use Questionnaire
N = 107
Questions: See text

PROFESSIONAL ADVANCEMENT: "MEANWHILE, BACK AT THE OFFICE . . ."

A separate class of items asks about professional advancement (Table 7.11). There is little difference in perceived impacts in the long term vs. the short term. But note an implicit tension between general scientific status and advancement within the specific organization by which one is employed. On the local organizational scene, one's connectivity to a national scientific network is apparently frequently perceived as damaging immediate advancement. Anecdotal evidence from users indicates that some employers deeply resent these organizationally external contacts and efforts, and occasionally even try to deprive the employee of access.

TABLE 7 · 11 ***Impacts on Professional Advancement***

Participation in EIES contributes to:

Short-term professional advancement in terms of my current employment

Strongly Agree	Agree	Neither Agree nor Disagree	Disagree	Strongly Disagree
6%	25%	30%	26%	15%

Short-term professional advancement in terms of my status among my peers in my specialty

Strongly Agree	Agree	Neither Agree nor Disagree	Disagree	Strongly Disagree
7%	35%	37%	13%	9%

Long-term professional advancement with respect to employment

Strongly Agree	Agree	Neither Agree nor Disagree	Disagree	Strongly Disagree
3%	29%	37%	19%	13%

Long-term professional advancement with respect to my status among my peers in my specialty

Strongly Agree	Agree	Neither Agree nor Disagree	Disagree	Strongly Disagree
7%	35%	39%	12%	8%

Remember that communication with co-workers in the traditional, four-walled office did not decrease as a result of system use (Table 7.6). However, the manager often seems to feel that time online somehow "takes away" from time available for work at the traditional office. She/he cannot see nor understand what the worker is doing all that time online. Unless one's supervisors are also participating in the network, a certain amount of suspicion of malingering and/or resentment of loss of control over the employee are bound to occur in a large proportion of cases.

MULTIVARIATE ANALYSIS

Using all of the factors combined into an overall productivity index, we can next see how our intervening variables—time on the system and increases in connectivity—relate to perceived increases in productivity. A stepwise multiple regression was done using the three variables identified which individually correlate most highly with both the items in the index and the productivity index as a whole. These are:

TIME ON = Cumulative number of hours online at post-use

COMM = "How many different people do you feel that you are actually exchanging information or communicating with on this system, currently?"

EIES MET = "Of these, how many have you 'met' (gotten to know) over EIES?"

The results are shown in Table 7.12. We see that the most important determinant of subjective judgments of a productivity increase as a result of EIES use is how many new people one is communicating with online whom one actually met through the system. Hours online and the total

TABLE 7 · 12 ***Determinants of Increased Productivity: A Stepwise Multiple Regression***

Factor	Mult R	R Square	Beta
EIES Met	.47	.22	.39
Time On	.54	.29	.25
# Comm	.54	.29	.05

$F = 13.2$, $p < .01$
N of cases = 101
Note: see text for definition of variables

number of persons whom one is communicating with also make significant contributions. Together, these three variables have a multiple correlation with perceived productivity increase factors of .54, meaning that they explain 29% of the variance. We do not have quantifiable measures of the changes in work patterns which were explored with open-ended questions and described above. If we did, and these were added to the equation, we would expect to be able to explain an even greater proportion of the variance.

VARIATIONS

Many of the scientists using EIES reported no significant increases in productivity. Initial (pre-use) individual differences in attitudes toward the system are good predictors of subsequent system use and of perceived benefits two years later. There were also variations by group—some of the scientific research communities were much more successful and satisfied with their online communications than others. Variations in the nature of the task which they were performing, their size, and the amount and quality of leadership which they received seem to account for these variations by group.

The data on improvements in quality and quantity of work collected for Group 80, the hepatitis knowledge base project, show a reversal of the pattern of answers for the other groups. Group 80, the only task related group, has more agreeing that EIES increases the quantity of work that they are able to accomplish (nine out of twelve, or 75%) than agreeing that it has improved the quality of their work (five out of twelve, or 42%).

One might hypothesize tentatively that task related groups are more likely to report overall gains in quantity of work than non-task groups. This speculation is supported by the findings of the study of the "Terminals For Managers" project at Stanford. Rice and Case (1981:9) report that after several months of use, an overall 45% of their senior level managers perceived increased quantities of work being accomplished due to system use, whereas a somewhat smaller 31% perceived improvements in quality of work. Replicating the findings of our study, they also report a very strong association between amount of use of the TFM message system and perceived improvements in productivity.

Inconsistent results among studies which have examined productivity increases for various systems and types of user groups suggest that system design, training, and user support services also influence productivity gains (See Kerr and Hiltz, 1982). Productivity enhancements are a potential of the medium, but are by no means always realized.

INTERVENING VARIABLES: A PATH MODEL

Our next step is to extend this analysis backwards to join with earlier analyses of determinants of amount of system use to model the entire process which occurs on EIES. A PATH analysis was used for this purpose (see Duncan, 1966, and Kim and Kohout, 1975). A series of univariate and multi-variate regression analyses are conducted to determine the strength of the relationships among the factors, which are shown in the diagram (Table 7.13) by the standardized regression coefficients (Beta).[1]

The model starts with our best predictors of EIES use. The variables in the middle of the model are hypothesized as intervening factors with both direct and indirect causal links to the Time 1 (pre-use) variables and the Time 3 (post-use) outcomes. For instance, estimated hours online has a weak but significant relationship with the number of people met on EIES. Time actually spent online has both a direct effect on increased productivity, and an indirect effect. Time online increases the number of persons met online, which in turn is our strongest predictor of productivity increases. The number of persons met on EIES also increases

TABLE 7 · 13 ***A PATH Diagram of EIES Use (Numbers shown are Beta Coefficients)***

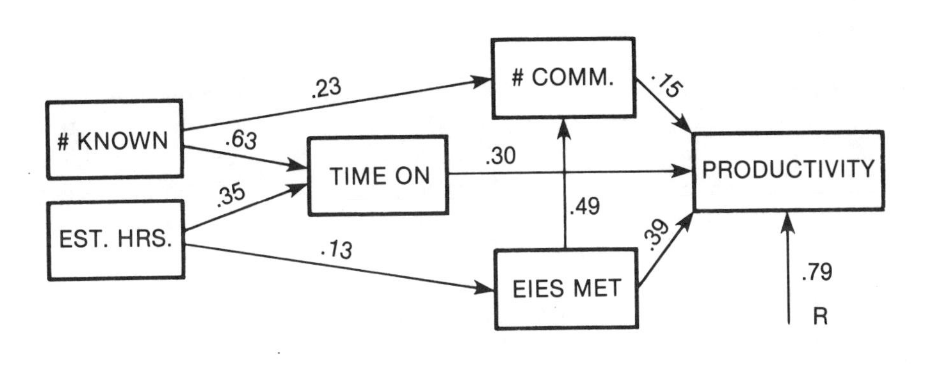

[1] It should be noted that this model was both the initial or theoretical model, and the "final" or empirically observed model; there was no fiddling with variables to make the results come out better.

the number of persons being communicated with online, which is another direct determinant of productivity.

Not all of the possible indirect links are shown, either because they are considered theoretically unimportant or unlikely and/or because empirically the causal link has no evidence. For instance, it might be thought that perhaps the number of persons met online is affected by the number known before use, since one could be introduced to new acquaintances by old ones. However, there is no significant relationship. Likewise, one might posit that time online alone increases the number of persons communicated with, directly. In fact, there is no significant relationship (Beta = .03).

One of the weakest links in the model is the determinants of the key variable "EIES MET" (number of persons with whom one is actively communicating who were "met" on the system). Like the initial level of estimated hours online, this is probably determined by a number of unmeasured motivations and personality factors. There are hundreds of potential new colleagues online in a system like EIES; all members have an equal opportunity to communicate with each other; yet some take advantage of this opportunity and some do not. Those who do meet many new people online are likely to be happier with the system and to perceive significant productivity increases in their work. What determines the number of people whom one will reach out to meet and communicate with on a system like this is a process worthy of detailed study.

SUMMARY AND CONCLUSIONS

In terms of effects on media use, EIES communication is most likely to be added onto other communications; but those who use the system the most are likely to also expand their use of other communications modes. There is some replacement of telephone and mails by computer-mediated communication. Travel to professional meetings and visits with other researchers are not affected for most people. Although the majority report no effect on the reading of professional books and journals, a significant minority (28% overall and 44% of heavy system users) report an increase. Communication with colleagues at one's own location is more likely to increase than to decrease.

Of course, subjective reports about the frequency of use of various media are likely to be quite unreliable. However, we did not ask for accurate counts, but only for gross changes: up, down, or about the same. Overall, there is a tendency for the media to add on to other modes and channels of communication, rather than to substitute for them. Previously established scientific and professional networks, maintained by other forms of communication, persist along side of the new, larger, more widespread computer-mediated network.

Nourished by this additional communication in a new form, various measures of social ties show strengthening. New ties are established on the computer network, and some of the new professional colleagues become personal friends as well as co-authors or collaborators. In social network terminology, the community becomes not only larger but more densely knit.

The majority of EIES users report some general impacts of the system on the way in which they think and work. Perspectives are broadened—with exposure to more ideas, theories, scientists, and opinions. Subjectively experienced effects of the increased communication with a larger network of scientists include the ability to get "instant feedback" on ideas and to "kick ideas around" with others when a piece of work is in its formative stage. They feel less constrained by the time and space which separates them from their colleagues. Finally, there is an adjustment in working patterns and habits—one to three hours a day online fitted into the schedule of heavy users; increases in the speed and pace of work; and in feelings of information overload and the need to organize one's work more formally.

Turning to productivity, as would be expected, the more time spent online, the more likely users are to subjectively report increases in the quantity and quality of work accomplished as a result of system use. Increases in quality of work are more likely to be perceived than increases in quantity (better reports or articles, rather than more). Such productivity increases seem to be linked to reported increases in the "stock of ideas" with which to attack new problems, and to the availability of leads, references and other information which can be used to help in one's work.

A multivariate analysis indicates that "meeting" new people online plays a central part in the process and outcome of system use. The strongest predictor of subjectively reported increases in productivity is the number of persons met on EIES with whom one subsequently establishes regular exchanges.

Perhaps EIES is like an intellectual lonely minds club or singles bar. People come to it hoping to expand their contacts, establish some "meaningful" communications, be stimulated by new ideas. If they do meet a lot of intellectually compatible people online, they feel that the experience is a productive one.

By no means all of the scientists who used EIES reported "meeting" new people or experiencing improvements in productivity-related factors. In addition, we have no basis on which to generalize the findings to other systems or applications. A synthesis of the findings of nine studies on a variey of systems and types of knowledge workers does show unanimous support for the finding that connectivity tends to increase; reported increases in the overall quality of work are not as consistent (Kerr and Hiltz, 1982).

In sum, the results of this case study do indicate that the use of a computer-mediated communication system can result in significant increases in productivity. Future research should focus upon identifying those conditions under which productivity increases are most likely to occur. There may be steps that can be taken in such areas as system design, user training and support, selection of appropriate tasks to be accomplished online, and effective managerial or leadership techniques in this medium that will significantly improve the percentage of cases in which the productivity potentials of the medium are actually realized.

ment. The
permits all
example, the data
views, and
to gross
support
of the new
expert
in a given
which
its belief. The
those different
to one view where the
one view. In some

8.

SUMMARY AND CONCLUSIONS

Computer-mediated communication systems appear to be the "cheapest and fastest" route to the "office of the future." But like the truck-filled routes of an eight-lane superhighway, they may not be the easiest route to negotiate for those who are used to the more leisurely pace of rambling two lane roads.

As Bair (1978:733) has pointed out in an article entitled "Communication in the Office-of-the-Future: Where the Real Payoff May Be",

> An examination of the resources of U.S. business shows that nonclerical personnel are the largest labor cost, and the principal activity of nonclerical staff is communication. Thus, the greatest leverage for the benefits of office automation is in supporting the communication activities of nonclerical personnel . . . If the behavioral problems can be overcome, computer mail could provide the highest payoff in office automation.

A later work expands upon the nature of these "behavioral" problems and the use of evaluation research to identify and deal with them:

> "The single most common cause of system failure is user rejection . . . The way the system is implemented has caused the most failures by not overcoming the threatening nature of the complex and intrusive technology . . . The principles that have been derived from research and experience should be followed to reduce the probability of rejection. Among those principles is ensuring the opportunity for user feedback to the system developers and management. (Uhlig, Farber, and Bair, 1979:243)

In this chapter, we will summarize the findings of this case study and their implications for overcoming the resistance of "knowledge workers" to using the computer to augment their communications. A "primer" on the basic strategies and guidelines for providing evaluation and feedback during system implementation is provided elsewhere (Kerr and Hiltz, 1982). However, we shall take a reflexive look at the methodological perspective that evolved during the several years the author spent living in and studying the natives in this electronic society.

SUMMARY OF FINDINGS AND THEIR IMPLICATIONS

ACCEPTANCE

The strongest predictors are attitudinal and motivational variables rather than any "objective" characteristics of users, such as previous computer experience or typing ability. Such variables include expectations about how useful the system will be and how many people one knows who are expected to be online.

Learning to use a new medium of communication and to effectively integrate it into one's work patterns is no simple matter. Although EIES members could learn the basics of using the system in a few hours, they did not become fully comfortable utilizing some of its potentially powerful features, such as joint document production, until 50 to 100 hours of experience.

Participants should feel that the task or activity in which they are engaging online is important enough so they are willing to *make time* to spend at least an hour a week online. Less regular participation leads to frustration for group members when messages are not picked up and responded to. Irregular participation is also associated with the user constantly forgetting how to use the system and never becoming proficient and comfortable with it.

With self-activated learning, as occurred with EIES, those with poor initial expectations of the usefulness and importance of communication with others via the computer system are likely to never sign on at all or to lack the motivation to remain through the learning period. The very high dropout rate among invited users is a serious problem for the future of computer-mediated communication systems. A supportive human facilitator (such as the EIES user consultants) seems to play a vital role in encouraging and helping new users through their crucial first few sessions.

A user group must include one or more persons willing to take the responsibility for an active leadership role. Effective leaders spend an hour or more a day online to organize and stimulate the interaction and task coordination. Managers or leaders must select appropriate applications and provide support for the users as they are learning to use the new medium.

THE EVOLUTION OF ATTITUDES AND BEHAVIOR

Those who do get through the learning period and actually participate in group communications tend to rate the system positively in terms of such characteristics as being easy, fun, and productive. They also tend to endorse specific design choices that were made in the EIES system, such as forced delivery of private messages (inability to reject them before they are ever printed out) and a progression of interface levels beginning with menus. The strongest predictor of subjective satisfaction with the system is the extent to which it has expanded social networks through facilitating "meeting" and working with new colleagues who share one's interests.

There is a process of "evolution" in user behavior. More experienced users change their preferred mode of interaction from passive menu selection to active command definition, expand and change the nature and number of system features which they consider necessary and useful, and expand the range of communications functions for which the medium is seen as satisfactory.

User group is an important contextual variable. The same system is likely to be perceived as having good or bad software features and as being a productive or an unproductive means of working with others, as a function of group membership. Group membership includes such variables as whether or not there is effective leadership and the nature of the task the group is working on.

"Design evolution" to support the acquired sophistication of experienced users appears to improve feelings that the system is "theirs," a tool responsive to their needs. Groups like to be able to participate in the development of some special structures or commands to help them in their particular tasks.

Even though there were many differences among groups, applications, and systems, results for several measures of subjective satisfaction replicated on EIES, MACC-TELEMAIL, PLANET, and NLS are for the most part very similar. This implies that there are some general characteristics of all computer-mediated communication systems in terms of user reactions.

IMPACTS

In examining whether the technology produces the intended increases in productivity of "knowledge workers," the key variable is not the attributes of the system itself. It is whether the technological potential for increasing social "connectivity" is indeed realized.

There were no dramatic "scientific revolutions" in the sense of new paradigms emerging during the eighteen to twenty-four months of observation. However, there was progress towards the clarification of theoretical controversies in most groups. Most importantly, the professional network of active users expanded; they had a feeling of greater awareness of varieties of work in the area and of the availability of new sources and types of information useful in scientific work. In addition to an increase in total communication within the specialty, there was also an increase in communication across disciplines or specialities. There were no decreases in communication with off-line colleagues as a result of system use.

Subjectively perceived increases in "quality" of work as a result of system use are more frequently reported than increases in quantity of work performed. Exposure to a broader range of information and ideas than otherwise possible, and the availability of a much larger network of people who may be helpful when one does want information and assistance with a specific project, are among the benefits that are seen as increasing productivity. In terms of media substitution, there is some decrease in telephone and mail use as a result of substituting computerized conferencing, but no decrease in travel or in reading of professional books and journals.

METHODOLOGICAL POSTSCRIPT MEANING AND OBJECTIVITY: PARTICIPANT OBSERVATION AND THE RAIN DANCE

Initially, this study was designed to measure some rather limited, predefined impacts of a computerized conferencing system on the communication patterns, paradigms, and productivity of scientific research communities. The process of communication via computer was seen as merely an intervening variable. As the study progressed and the communication patterns were observed, however, the process of communication via computer emerged as a phenomenon worthy of study and description in its own right. The study thus shifted in focus as it progressed, from the sociology of science to the sociology of individual and group processes in adapting to a new communications medium. In addition, it became obvious that this new form of communication had some unanticipated consequences for the participants. Even though some of the scientists wondered if the amount of time they invested in such communication had any direct productivity payoffs, they continued to participate. At the same time, the detached observer became caught up in a shared belief by the members of the communities that their activities had some importance and significance for the future of scientific research, and society as a whole, even if they were not quite sure how to explain the significance.

There are parallels with the cautionary tales of Paul Lazarsfeld as he discussed the dangers of "going native" while studying the rain dance, and with the penetrating functional analysis of Robert Merton in illustrating the concept of latent functions with that very same ritual gathering. Lazarsfeld cautioned his students somewhat as follows:

> You can observe the rain dance and maintain your objectivity. You can even participate in the rain dance, and gain a subjective understanding of its meaning for participants. But when you start to believe in the rain dance—when you start scanning the horizon anxiously for sight of those dark clouds signalling that your activity is indeed going to bring rain—then you are in trouble. You have gone native.

As this study progressed, I did begin to share the belief of many of the participants that their experiences with a new technology had important consequences for the future of not only scientifiic communities and other knowledge workers, but also human society in general. This belief could not be substantiated with any objective evidence of productivity gains. There are only the subjective feelings of the most active participants that their electronic tribal gathering was beneficial; that if nothing else, they enjoyed it and were stimulated by it.

From the point of view of traditional functional analysis, the premise is that if a group or society persists in a pattern of behavior, then it must have some beneficial outcomes. These outcomes may be "latent functions" -neither intended nor necessarily recognized by the participants, as contrasted to the "manifest functions," those publicly announced, officially endorsed goals for the activity (Merton, 1968, p. 119). In the case of the rain dance, there is no objective evidence that the amount or form of dancing affects the probability of rainfall. A parallel that struck me is that an important scientific breakthrough may be just as unpredictable as rain. Hard work alone may not cause it to happen; other environmental variables are a controlling factor. However, in the tradition of Durkheim (in *The Elementary Forms of the Religious Life*), the rain dance may have important functions for the social solidarity of the group. It brings the members of the tribe together, gives them a chance to reaffirm and clarify their shared beliefs, and creates the opportunity for new marriages and alliances. Such new alliances may indeed help the participants to better cope with their environment and engage in fruitful cooperative efforts, over the long run.

For the new class of computer-based communications systems, of which EIES is a forerunner, but only a single example, it is hard to quantify those latent functions, which may be the most important in the long run. In retrospect, a period of eighteen to twenty-four months seems too short to expect to see large increases in the productivity of individual participants or dramatic paradigm shifts in the scientific communities. EIES activity was, after all, only a small portion of their total professional lives and activities. For the relatively "heavy" users who spent 100 or more hours online over eighteen to twenty-four months, this is still only a few hours a week. The most important consequences seem to be the enlarged and strengthened professional network of colleagues, the greater understanding of the work of others and how it relates to one's own, and the feelings of having been exposed to information and ideas that form a permanent resource for the rest of one's professional life.

Describing and understanding the communication activity itself, its forms and variations, and the feelings of the participants about it, became an emergent objective of the study. As in an ethnographic study, one can gather some qualitative descriptive data by observing and participating, and some more quantitative descriptive data by surveys of attitudes and census counts of activity patterns. When we have a number of such descriptions for various computer-based systems and a variety of user communities, we will be in a better position to try to prove "cause" and "effect."

Much early work in anthropology fell into the category of "ethnography": the description of a single society. Later, as this descriptive material accumulated, "ethnology," or the comparison of similar institutions across societies, became possible. A priority for future research on computer-mediated communication systems should be sufficient standardization of the types of data collected and the measurement instruments used so that an "ethnology" of computer-mediated social systems becomes possible. We need to discover how the kinds of results observed for this case study may vary for other kinds of knowledge workers and for other systems with a different mix of features.

In the meantime, the forms and rituals of communication via computer are at least as interesting to study as the rain dance, and potentially much more important for the future of a society which may be forced to choose cheap telecommunication alternatives in an era of scarce resources. The rain dance may be seen as a cultural reaction to a crisis situation with which the society does not know how to deal. Experimentation with new computer and communication technologies may represent attempts to deal with the problems of post-industrial societies.

CONCLUSION

Looking at the high dropout rate contrasted with the testimonials of the confirmed users, one wonders if perhaps Computerized Conferencing is like religion: it only helps if you have faith that it will.

This case study illustrates the complex interdependence of technological potentials and social structural variables in determining the success of technological innovations. Computer-mediated communication systems are not a technological magic wand that can be waved over an organization to achieve instantaneous transformations in productivity. It takes some time for new users to become comfortable with the medium and realize the potentials that it offers. It also takes the right "social implementation," from the initial choice of an application that will supply a "critical mass" for the online community, through constant attention to facilitating or managing the group's work online. With the necessary attention to social engineering, the technology can improve the effectiveness of professional and technical workers in the "offices" of the future. In doing so, however, it is likely to profoundly change the nature of white collar work and of the organizations and communities which use it.

REFERENCES

Academic Computing Center, The University of Wisconsin-Madison
1975 EDIT. Reference Manual for the 1110. Revision 2.
1977 MACC-TELEMAIL System. User Manual for Univac 1100 Series Computers.

Adriansson, Lillemor, and Sjoberg, Lennart
1980 "Gruppkommunikation via dator: Inledande Social-psykolgiska Studier pa KOM-Systemet vid FOA." Rapprot 2. Gothenburg University: FOA rapport C 560–24–H9(M3).

Allison, Paul D., and Stewart, John A.
1974 "Productivity Differences Among Scientists: Evidence for Accumulative Advantage." American Sociological Review, 39, 4 (August) 596–606.

Bair, James H.
1978 "Communication in the Office of the Future: Where the Real Payoff May Be." Proceedings of the International Conference on Computer Communications, Kyoto, Japan, 733–739.

Bell, Daniel
1973 The Coming of Post-Industrial Society: A Venture in Social Forecasting. New York: Basic Books, paperback edition.

Bennett, J.L.
1972 "The User Interface in Interactive Systems." Annual Review of Information Science and Technology, Vol. 7, 159–196.

Bureau of the Census, U.S. Dept. of Commerce
1980 Statistical Abstract of the U.S. Washington, D.C.: U.S. Govt. Printing Office.

Caldwell, Roger
1981 "Conference Management Case History: Energy and Environment." EIES Paper Fair (C1017cc97), April 15, 1981.

Carey, John
1980 "Paralanguage in Computer Mediated Communication." Proceedings of the Association for Computational Linguistics, 61–63.

Chubin, Daryl E.
1975 "On the Conceptualization of Scientific Specialties: The Interplay of Demography and Cognition," Paper presented at the 1975 annual meeting of the American Sociological Association.
Cole, Jonathan R., and Cole, Stephen
1973 Social Stratification in Science. Chicago IL: University of Chicago Press.
Coleman, J., Katz, E., and Menzel, H.
1966 Medical Innovation: A Diffusion Study. New York: Bobbs-Merrill.
Crane, Diana
1972 Invisible Colleges: Diffusion of Knowledge in Scientific Communities. Chicago IL: University of Chicago Press.
Crittenden, K.S. and Montgomery, A.C.
1980 "A System of Paired Asymmetric Measures for Use with Ordinal Dependent Variables." Social Forces 58,4 (June):1178–1186.
Duncan, Otis D.
1966 "Path Analysis: Sociological Examples." American Journal of Sociology, 72: 1–16.
Durkheim, Emile
1961 The Elementary Forms of the Religious Life. New York: Collier Books.
Edwards, Gwen
1977 An Analysis of Usage and Related Perceptions of NLS—A Computer Based Text Processing and Communications System. Montreal: Bell Canada.
Elton, Martin
1974 "Evaluation of Telecommunications: A Discussion Paper." London: Communications Studies Group, Report P/74244/ST/.
Freeman, Linton, and Freeman, Sue
1980 "A Semi-Visible College: Structural Effects on a Social Networks Group," in M.M. Henderson and M.J. MacNaughton, eds., Electronic Communication: Technology and Impacts. AAAS Selected Symposium 52. Boulder, CO: Westview Press.
Garvey, William D., and Griffith, Belver C.
1971 "Scientific Communication: Its Role in the Conduct of Research and Creation of Knowledge." American Psychologist, 26, 4 (April): 349–362.
Griffith, Belver C. and Mullins, Nicholas C.
1972 "Coherent Groups in Scientific Change: 'Invisible Colleges' May Be Consistent Throughout Science." Science 177: 959–964.
Guillaume, J.
1980 "Computer Conferencing and the Development of an Electronic Journal," Canadian Journal of Information Science, 21–29.
Hagstrom, Warren D.
1965 The Scientific Community. New York: Basic Books.
1970 "Factors Related to the Use of Different Modes of Publication. Research in Four Scientific Fields." pp. 85–124 in C. E. Nelson and D. K. Pollack, eds. Communication among Scientists and Engineers. Lexington, MA: Lexington Books.
1976 "The Production of Culture in Science." American Behavioral Scientist, 19, 6: 753–767.
Heller, S., Milne, G.W.A., and Feldman, R.J.

1977 "A Computer Based Chemical Information System." Science 195, (21 Jan):253–259.

Hiltz, Starr Roxanne

1979 "Using Computerized Conferencing to Conduct Opinion Research." Public Opinion Quarterly, Winter.

1981 The Impact of a Computerized Conferencing System on Scientific Research Communities. Newark, N.J.: Computerized Conferencing and Communications Center, New Jersey Institute of Technology, Research Report No. 15.

Hiltz, Starr Roxanne, Johnson, Kenneth, Aronovitch, Charles, and Turoff, Murray

1980 Face-to-Face vs. Computerized Conferences: A Controlled Experiment. Newark, NJ: Computerized Conferencing and Communications Center, New Jersey Institute of Technology, Research Report Number 12.

Hiltz, Starr Roxanne, Johnson, Kenneth, and Turoff, Murray

1982 The Effects of Formal Leadership and Computer Feedback on Group Decision Making Via Computer: A Controlled Field Experiment, Research Report Number 18. Newark, NJ: Computerized Conferencing and Communications Center, NJIT.

Hiltz, Starr Roxanne, and Kerr, Elaine B.

1981 Studies of Computer-Mediated Communication Systems: A Synthesis of the Findings. Final Report to the National Science Foundation.

Hiltz, Starr Roxanne, and Turoff, Murray

1978 The Network Nation: Human Communication via Computer. Reading, MA: Addison Wesley Advanced Book Program.

1982 "Office Augmentation Systems: The Case for Evolutionary Design." Paper presented at the Fifteenth Hawaii International Conference on System Sciences.

Johansen, Robert

1976 "Pitfalls in the Social Evaluation of Teleconferencing Media." In Horne, A.P. and Riccomeni, B. eds., The Status of the Telephone in Education. University of Wisconsin Extension Press.

Johansen, Robert, DeGrasse, Jr., Robert, and Wilson, Thaddeus

1978 Group Communication Through Computers, Vol. 5, Effects on Working Patterns. Menlo Park, Cal., Institute for the Future, Report R–41.

Johansen, Robert, Vallee, Jacques, and Spangler, Kathleen

1979 Electronic Meetings: Technical Alternatives and Social Choices. Reading, MA.: Addison Wesley.

Johnson-Lenz, Peter and Trudy

1981a The Evolution of a Tailored Communications Structure: The Topics System. Newark, NJ: Computerized Conferencing and Communications Center, NJIT, Research Report Number 14.

1981b "Consider the Groupware: Design and Group Process Impacts on Communication in the Electronic Medium," in Hiltz and Kerr, 1981.

Kerr, Elaine

1980 "Conferencing Via Computer: Evaluation of Computer-Assisted Planning and Management for the White House Conference on Library and Information Services." In Information for the 1980s: A Final Report of the White House Conference on Library and Information

Services, 1979, 767–805. Washington, DC: U.S. Government Printing Office.
Kerr, Elaine B. and Hiltz, Starr Roxanne
1982 Computer-Mediated Communication Systems: Status and Evaluation. New York: Academic Press.
Kincaid, D.
1979 "Communication Network Analysis and Innovations in Rural Villages." Presentation to the Seminar on Communication Network Analysis, East-West Center, Honolulu, Hawaii.
Kollen, James, and Garwood, John
1975 Travel/Communication Tradeoffs: The Potential For Substitution among Business Travellers. Report published by Bell Canada Business Planning Group.
Kim, Jae-On
1975 "Factor Analysis," in Norman Nie et al., SPSS Statistical Package for the Social Sciences, second edition: 468–514. New York: McGraw Hill.
Kim, Jae-On and Kohout, Frank J.
1975 "Special Topics in General Linear Models." In Norman Nie et al., SPSS Statistical Package for the Social Sciences, second edition: 368–397. New York: McGraw Hill.
Kuhn, Thomas S.
1970 The Structure of Scientific Disciplines, Revised Edition. Chicago IL: The University of Chicago Press.
Landweber, Lawrence H.
1977 "An Electronic Mail-Box and Teleconferencing Network for Theoretical Computer Science." Proposal to the National Science Foundation.
Leven, R., and Schroeder, M.D.
1979 "Transport of Electronic Messages Through a Network." In E.J. Boutmy and A. Danthine, eds., Proceedings, Teleinformatics. Amsterdam: North Holland Press: 29–33.
Lipinski, Hubert, and Miller, Richard H.
1974 "FORUM: A Computer Assisted Communications Medium." Proceedings, Second International Conference on Computer Communications. Stockholm: 143–147.
Lucas, Henry C., Jr.
1975 Why Information Systems Fail. New York: Columbia University Press.
Markle, Gerald E., and Fox, John W.
1975 "Paradigms or Public Relations: The Case of Social Biology." Paper presented at the 1975 meetings of the ASA.
Martino, Joseph
1977 A Computer Conference on Futures Research. Proposal submitted to the National Science Foundation.
Martino, Joseph, and Bregenzer, John
1980 Report on an Experiment with an Electronic Conferencing System within a Scientific Community. Final Report to the National Science Foundation.
McCarroll, Jane
1980 "EIES for a Community Involved in R&D for the Disabled," in M.M. Henderson and M.J. MacNaughton, eds., Electronic Communication: Technology and Impacts. AAAS Selected Symposium 52, Boulder, CO: Westview Press: 71–76.

Merton, Robert K.
1968 Social Theory and Social Structure. Enlarged Edition. New York: Free Press.
1968 "The Matthew Effect in Science." Science, 59(Jan) 56–63.
1973 The Sociology of Science. Chicago, IL: The University of Chicago Press, 1973.
Mintzberg, Henry
1973 The Nature of Managerial Work. New York: Harper and Row.
Mitroff, Ian A.
1974a The Subjective Side of Science. Amsterdam: Elsevier.
1974b "Norms and Counter-Norms in a Select Group of the Apollo Moon Scientists: A Case Study of the Ambivalence of Scientists." American Sociological Review, 1974, 34, 4: 579–595.
Mulkay, Michael
1972 "Conformity and Innovation in Science." P. Halmos, ed. The Sociological Review, Monograph 18: 5–23.
1976 "The Mediating Role of the Scientific Elite." Social Studies of Science, 6, 3–4 (Sept. 1976), 445–470.
Mullins, Nicholas C.
1972 "The Development of a Scientific Specialty: The Phage Group and the Origins of Molecular Biology." Minerva 10 (January): 51–82.
Nilles, Jack M., Carlson, F.R., Jr., Gray, P., and Hanneman, G.
1976 The Telecommunications-Transportation Tradeoff: Options for Tomorrow. New York: Wiley.
NSF76–45
1976 National Science Foundation, Division of Science Information, "Program Announcement: Operational Trials of Electronic Information Exchange for Small Research Communities"
Oreskes, Michael
1981 "Productivity: The Search for the Old Magic." The New York Times, October 11, Section 12, p. 6.
Palme, Jacob
1981a Experience with the Use of the COM Computerized Conferencing System. Draft report, Stockholm, Sweden: Swedish National Defense Research Institute.
1981b Experience with the Use of the COM Computerized Conferencing System. Stockholm: Swedish National Defense Research Institute (FOA), Report C 10166E–M6(H9), December.
Panko, Raymond R.
1977 "The Outlook for Computer Mail." Telecommunications Policy (June).
Panko, Raymond R. and Panko, Rosemarie U.
1981 "A Survey of EMS Users at DARCOM," Computer Networks, 5: 19–23.
Parsons, Talcott
1951 The Social System. New York: The Free Press.
Price, Derek J. De Solla
1963 "Communication in Science: The Ends—Philosophy and Forecast." In CIBA Foundation, Communication in Science, 199–213.
Price, Derek J. De Solla and Donald De B. Beaver
1966 "Collaboration in an Invisible College." The American Psychologist 21 (Nov., 1966). pp. 1011–18.

Renner, R.L., Bechtold, R.M., Clark, C.W., Marbray, D.D., Wynn, R.L., and Goldstein, Nancy
1972 "EMISARI: A Management Information System Designed to Aid and Involve People." Proceedings Fourth International Symposium on Computers and Information Systems (COINS IV), Plenum Press.
Reskin, Barbara
1977 "Scientific Productivity and the Reward Structure of Science." American Sociological Review, 42 (June): 491–504.
Rhodes, Sarah N. and Bamford, Harold E. Jr.
1976 "Editorial Processing Center: A Progress Report." The American Sociologist, 11 (August): 153–159.
Rice, Ronald E.
1982 Human Communication Networking in a Teleconferencing Environment. Ph. D. Dissertation, Department of Communication, Stanford University.
Rice, Ronald E., and Case, Donald
1981 "Electronic Messaging in the University Organization," Paper presented to the Speech Communication Association, Anaheim, California, November.
Robbins, David, and Johnson, Ron
1976 "The Role of Cognitive and Occupational Differentration in Scientific Controversies." Social Studies of Science: 6, 3–4 (Sept. 1976), 349–368.
Rogers, E., and Kincaid, L.
1981 Communication Networks: Toward a New Paradigm for Research. New York: Free Press.
Sheridan, T., Senders, J., Moray, N., Stoklosa, J., Guillaume, J., and Makepeace, D.
1981 Experimentation with a Multi-Disciplinary Teleconference and Electronic Journal on Mental Workload. Final Report to the National Science Foundation. Cambridge, Mass.: Massachusetts Institute of Technology.
Short, John, Williams, Ederyn, and Christie, Bruce
1976 The Social Psychology of Telecommunications. New York: Wiley.
Siegel, E.R.
1980 "Use of Computer Conferencing to Validate and Update NLM's Hepatitus Data Base." In M.M. Henderson and M.J. MacNaughton, eds., Electronic Communication: Technology and Impacts. AAAS Selected Symposium 52, Boulder, CO: Westview Press, 87–95.
Stevens, Chandler Harrison
1980 "Many-to-Many Communication through Inquiry Networking." World Future Society Bulletin, Vol. 14: 31–35.
Storer, N. W.
1966 The Social System of Science. New York: Holt Rinehardt and Winston.
Turoff, Murray
1972 "Party-Line and Discussion: Computerized Conferencing Systems." Proceedings, First International Conference on Computer Communications, Wash. D.C., 161–171.
Turoff, Murray, and Starr Roxanne Hiltz
1978 Development and Field Testing of an Electronic Information Exchange System: Final Report on the EIES Development Project. Re-

search Report Number Nine: Computerized Conferencing and Communications Center New Jersey Institute of Technology.

Uhlig, R.P.
1977 "Human Factors in Computer Message Systems." Datamation (May): 120–126.

Uhlig, Ronald P., Farber, David J., and Bair, James H.
1979 The Office of the Future: Communication and Computers. Amsterdam: North-Holland Publishing Co.

Umpleby, Stuart
1977 "General Systems Theory: An Example of the Integration of Scientific Disciplines." Proposal submitted to the National Science Foundation.
1980 "Computer Conference on "General Systems Theory: One Year's Experience," in M.M. Henderson and M.J. MacNaughton, eds., Electronic Communication: Technology and Impacts. AAAS Selected Symposium 52, Boulder, CO: Westview Press, 55–63.

Vallee, Jacques, Johansen, Robert, Lupinski, Hubert, Spangler, Kathleen, and Wilson, Thaddeus
1975 Group Communication Through Computers, Volume 3: Pragmatics and Dynamics. Menlo Park, CA, Institute for the Future, Report R–35.

Vallee, Jacques, Johansen, Robert, Randolph, Robert, and Hastings, Arthur
1974 Group Communication Through Computers: Volume 2, A Study of Social Effects. Menlo Park, CA, Institute for the Future, R–33.

Vallee, Jacques, Johansen, Robert, Lipinski, Hubert and Wilson, T.
1978 Group Communication Through Computers, Vol. IV: Social, Managerial, and Economic Issues. Menlo Park, CA: Institute for the Future.

Whyte, William Foote
1980 "Exploring the Frontiers of the Possible: Social Inventions for Solving Human Problems." Announcement of 1981 program theme in Footnotes, Vol. 8. Washington, D.C.: American Sociological Assn.
1982 "Social Inventions for Solving Human Problems." American Sociological Review, 47,1: 1–13.

Wilcox, R., and Kupperman, R.
1972 "An On-Line Management System in a Dynamic Environment." Proceedings, First International Conference on Computer Communications, 117–120.

Zinn, K., Parnes, R., and Hench, H.
1976 "Computer Based Educational Communications at the University of Michigan." Proceedings 31st ACM National Conference.

APPENDICES

APPENDIX A
PRE-USE QUESTIONNAIRES AND MARGINALS, EIES

PRE-USE QUESTIONNAIRE
Study of the Impact of Computerized Conferencing
Upon Research Communities
(Copyright, 1977, Starr Roxanne Hiltz)

Your cooperation in completing the following questionnaire, before you participate in the system for more than an hour or so, is vitally necessary for a thorough and proper evaluation. The questions are designed to collect some information on your general background, your communication skills and style, your access to the conferencing system and your predisposition concerning its use. You should be able to complete the answers in about 30 minutes.

<u>Directions</u>

Most of the questions are structured so that they require only a check or a simple numeric response. Some, however, request you to list or describe items. Please type or print your response as clearly as possible. Where you do not know or cannot make a rough estimate of the answer you may leave it blank.

Notice that a continuation page has been attached to the end of the questionnaire should you need additional space to answer or clarify your response to any of these questions.

Your Name ______________________

EIES Group Name/# ______________________

Job Title ______________________

Your Employer ______________________

City ____________ State ________

This questionnaire is voluntary and in no way conditions your participation in the system. If you have, for some reason, an objection to filling out this questionnaire, please note your objection below and return it to us. Or, if the case applies, note your objection to any single question and leave it blank.

Objection:

PRE-USE QUESTIONNAIRE

DATE FILLED IN __________

CODED ID ONLY ___________

(Cover page will be removed to
preserve confidentiality)

TURN PAGE TO BEGIN

Part I. Your EIES Group's Research Specialty (Group # ____________)

1. Please give a one sentence description in your own words of the scientific or technical specialty of your EIES USER Group. (Note: this name will subsequently be what is meant by "your specialty area"). Then describe the main problem or project on which you personally are working, within this specialty area.

Employer =

Academic	71
Govt.	4
Priv. Research	9
Business	2
Medical	3
	89

1 missing

2. What is the approximate year in which this specialty became recognized (or will become recognized) as a separate and distinct research area?

5 = 10 5-9 = 18 10-19 = 18 20+ = 26 Tot. = 80

3. For approximately how long have you been actively working within this specialty area? ____________

1 = 2 1-4 = 20 5-9 = 35 10+ = 29 Tot. = 86

4. What is the total number of journals in which articles relevant to your specialty area are likely to appear?

(1)______ none
(2)__3__ two or less
(3)__49__ 3 - 10
(4)__15__ 11 - 19
(5)__7__ 20 - 49
(6)__4__ 50 - 99
(7)__7__ 100 or more

Tot. Ans. 85

5. Is there any journal or newsletter or other published source in which you can find descriptions of current (unfinished) research activities and developments within your specialty?

(1)__27__ No

Tot. Ans.

(2)__54__ Yes: please list:__81__

6. Is there any one meeting or convention which you "must" attend in order to keep up with research in your specialty? (IF yes, please list).

(1) 57 No (2) 29 Yes (Tot. 86)

7. Could you list the four major or outstanding people in your entire specialty and the extent to which you know them personally and/or are in direct contact with them?

	On EIES	Not on	Tot.	Constant	Frequently	Occasionally	Rarely	Never	Tot.
				Extent of Current Contact					
a.	28	42	70	1 = 12	2 = 21	3=22	4 =14	5 =6	75
b.	24	46	70	1 = 13	2 = 12	3 = 22	4 = 18	5 = 9	74
c.	25	40	65	1 = 6	2 = 13	3 = 23	4 = 21	5 = 7	70
d.	20	36	56	1 = 4	2 = 8	3 = 26	4 =11	5 = 12	61

8. How many members of your EIES User Group do you know either professionally or personally? Tot. 213

1-5 = 153 6-10 = 27 11-20 = 19 21-79 = 9 All = 3 Most = 2

9. Is there a commonly accepted "intellectual mainstream" in your specialty?

(1) 36 Yes (2) 49 No Tot. = 85

10. If yes; to what extent do you feel that you and those with whom you collaborate are in the recognized intellectual "mainstream" of your specialty, or conversely feel you are "isolated" or "peripheral"? (circle one)

Completely in the Mainstream	Somewhat in the Mainstream	Neither in the Mainstream nor Isolated	Somewhat Isolated	Completely Isolated	
1	2	3	4	5	
15	14	13	6		Tot. = 48

11. How would you rate the degree or intensity of competition within your research specialty?

recode

Very Intense	Intense	Moderate	Low	Nonexistent	
1	2	3	4	5	
17		42	23		Tot. = 62

12. What are the reasons for this competition? (Check all that apply).

yes	No		Tot.
yes = 34	No= 39	Scarcity of or competition for funds	73
21	52	Rival groups of collaborators	
38	35	High achievement or success drive of persons in the field	
7	66	Some persons act unethically	
21	52	Strongly opposing views	
11	62	Other (please describe) :	

13. Please list the name of any other research specialties in which you are currently involved, and whether you are currently spending more time or less time on each one than on your EIES specialty.

Name		Other or Equ.	More time	Less time	Tot.
Spec. 1	None = 14	4	41	30	= 89
Spec. 2	39	2	22	25	= 88
Spec. 3	65	4	7	12	= 88

Spec. Importance (Scale 0-6)

0 = 7
1 = 6

2 = 15
3 = 11
4 = 22

5 = 14
6 = 13

88

1. During an average week, approximately how many hours do you spend on each of the following kinds of activities? (First list the total for all professional activities, then the number of these related only to activities within your specialty area).

	Total	Hours in Specialty only
Direct research activities		
Writing papers, books, etc.		
Education		
teaching		
learning: reading books or journals		
attending meetings, seminars, etc.		
Administrative and support activities (committee meetings, memos, etc.)		
Telephone		
inside your organization		
outside your organization		
Consulting		
Funding (grants applications or other resource acquisition activities)		
Other professional activities (please specify)		
Total		

% Spec. Imp.

6 = 12
6-10 = 4
11-19 = 6
20-49 = 29
50+ = 31
82

2. Please list the names of any persons with whom you have co-authored or collaborated in research during the last year, or during the current one

0 = 16 3 = 16 Tot. 87
1 = 9 4-9 = 31
2 = 11 10+ = 4

3. Considering all current personal communication modes, what is the total number of different individuals within your research specialty with whom you are currently in contact? 0 = 3
1-2 = 9 3-5 = 11 6-9 = 6 10-19 = 25 20-49 = 15 50+ = 17 Tot. 86
4. How many of these are in your EIES user group? 0 = 9
1-2 = 24 3-5 = 22 6-9 = 10 10+ = 16 Tot. 81 Tot. 81
5. Scientists are sometimes anticipated by others in the presentation of research findings. That is, after they have started work on a problem another scientist publishes its solution. How often has this happened to you in your career? (Please exclude cases where a solution to your problem was published before you started your own work. Circle one.)

Recoded Constantly	Frequently	Time to Time	Rarely	Never	
1	2	3	4	5	
2	30	61	31	28	Tot. 90

6. How concerned are you that you might be anticipated in your current work?

Constantly	Frequently	Time to Time	Rarely	Never	
1	2	3	4	5	Tot. 90
9		58		23	

General Principles of Science

Described below are two sets of conflicting general principles which can guide the conduct and evaluation of scientific research. Please read each set of principles with your specialty area in mind.

Principle A. Emotional Neutrality

Scientists must be emotionally neutral and impartial towards their ideas if they are to stand a fair chance of ultimately being proved valid. Conducting an investigation with anything less than an impartial frame of mind possesses the danger that the scientist will bias results and be unable to give up hypotheses when they are indeed false.

Principle B. Emotional Commitment

Scientists must be emotionally committed to their ideas if they are to stand a fair chance of ultimately being proved valid. Unless a scientist believes intensely in his or her own ideas and does everything legitimately in his power to verify them, there is the danger that he will give up his ideas too quickly. Initial inconclusive signs of negative evidence do not warrant a reorientation of research efforts. The scientist must believe in himself and his own findings with great conviction.

7. On the basis of your own experience and observations, to what extent does each of the principles tend to govern the everyday working behavior of most scientists in your specialty? (Please circle one number).

<table>
<tr><td>Recoded
Tot. 85</td><td>A
Significantly More Than B</td><td>A
Moderately More Than B</td><td>Both Tend to Govern Equally</td><td>B
Moderately More Than A</td><td>B
Significantly More Than A</td><td>Neither Tends to Govern</td></tr>
<tr><td></td><td>1</td><td>2</td><td>3</td><td>4</td><td>5</td><td>6</td></tr>
<tr><td></td><td colspan="2">18</td><td>14</td><td colspan="2">53</td><td></td></tr>
</table>

8. To what extent does each of these principles tend to govern your own everyday working behavior?

<table>
<tr><td>Recoded
Tot. 89</td><td>A
Significantly More Than B</td><td>A
Moderately More Than B</td><td>Both Tend to Govern Equally</td><td>B
Moderately More Than A</td><td>B
Significantly More Than A</td><td>Neither Tends to Govern</td></tr>
<tr><td></td><td>1</td><td>2</td><td>3</td><td>4</td><td>5</td><td>6</td></tr>
<tr><td></td><td colspan="2">24</td><td>17</td><td>19</td><td>26</td><td>3</td></tr>
</table>

9. To what extent do you believe that each of the principles <u>ought</u> to govern the behavior of scientists in your specialty?

Recoded	A Signif- icantly More Than B	A Moder- ately More Than B	Both Equally	B Moder- ately More Than A	B Signif- icantly More Than A	Neither Should Govern
	1	2	3	4	5	6
	27 27		23	20	14	3

Principle C: <u>The Irrelevancy of Personal Attributes</u>

The personal attributes of a scientist are completely irrelevant in judging results and claims to knowledge. Each claim in science is judged impartially on its own merits by its ability to stand up to rational, empirical test procedures without reference to the particular scientist.

Principle D: <u>The Relevancy of Personal Attributes</u>

The personal attributes of a scientist are highly relevant in judging results and claims to knowledge. In reality the work of some scientists is given credence over that of others. It is necessary to know the personal characteristics, background and motivations of a scientist before one can properly evaluate his or her work.

As above, we wish you to indicate the extent to which these two principles tend to govern the everyday working behavior of <u>most scientists in your specialty</u>; tend to govern <u>your own</u> everyday working behavior, and ought to <u>govern</u> the behavior of scientists in your specialty.

Recoded Tot.		C Signif- icantly More Than D	C Moder- ately More Than D	Both Equally	D Moder- ately More Than C	D Signif- icantly More Than C	Neither
85	10. <u>Most scientists</u>	1	2	3	4	5	6
		27		11	47		
89	11. Your own behavior	1	2	3	4	5	6
		15	22	12	40		
88	12. Ought to govern	1	2	3	4	5	6
		27	20	19	21		1

Part III Background Items (Please attach a vita, if available; and omit items covered in the vita).

Tot.

1. What is your age?

88

(1) 2 under 25
(2) 28 25 - 34
(3) 38 35 - 44
(4) 16 45 - 54
(5) 3 55 - 64
(6) 1 65 & over

87 2. Sex: (1) 5 female (2) 82 male

3. Please list your academic degrees (Degree, Subject, Institution, and year).

76

Bach = 2 Masters = 10 No Degree = 1 Tot. 73

5 yrs = 16

Bach = 5 Ph. D, MD =59 5-9 = 29

10-19 = 18

20+ = 9

4. Have you ever won a prize, special award, or been elected to an honorary scientific society for your research accomplishments?

81

(1) 46 no
(2) 35 yes (Please list) ____________________

____

5. Professional Publications (please try to give exact numbers published in last year or underway; estimates are fine for previous works.)

81

	Currently in Progress	Published in Last Year	Published or Presented during Total Professional Career
Text books			
Other books			
Journal articles			
Papers presented			
Other (describe)			

6. I am more interested in generating a large number of alternate explanations for any problem than in pursuing one exclusively in detail.

90

Strongly Agree 1	Agree 2	Neither Agree nor Disagree 3	Disagree 4	Strongly Disagree 5
14	22	30	20	4

7. I prefer to work in well-established research areas.

Tot.	Strongly Agree 1	Agree 2	Neither Agree nor Disagree 3	Disagree 4	Strongly Disagree 5
92		2	31	49	10

8. How well known is your work, within your specialty area?

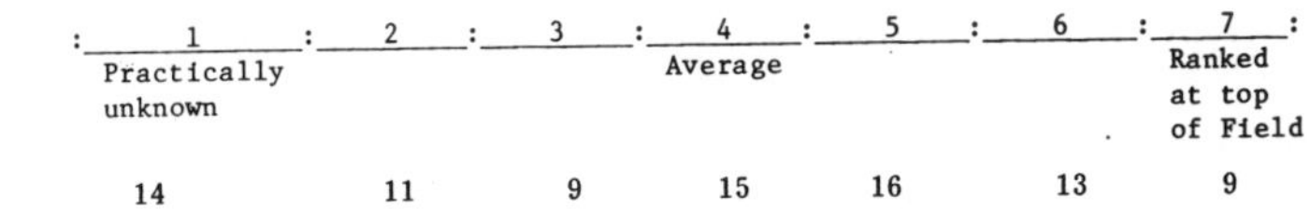

Tot.	1	2	3	4	5	6	7
87	14	11	9	15	16	13	9

Comments:

9. Do you think that the EIES system will affect familiarity with or the assessment of your work? Explain.

75

Yes, Considerably = 23

Yes = 26

Maybe = 14

No = 12

IV Communication Skills and Facilities

Tot.

1. Is English your primary language?

92

(1) 85 Yes (skip to question 2)
(2) 7 No

If not, what is your first language? ______

If English is not your first language, do you consider your English to be on a par with your primary language as to;

Writing	(1) 7	Yes	(2) 2	No
Speaking	(1) ___	Yes	(2) ___	No
Reading	(1) 8	Yes	(2) 2	No

2. How would you describe your English reading speed?

89

(1) 17 Very fast
(2) 54 Fast
(3) 18 Slow
(4) ___ Very slow

3. Comparing your writing skills and your speaking skills, would you say you were more persuasive when

89

(1) 43 Writing (2) 36 Speaking equal 10

4. How would you describe your typing skills?

91

(1) 3 None
(2) 19 Hunt and peck
(3) 35 Casual (rough draft with errors)
(4) 22 Good (can do 25 w.p.m. error free)
(5) 12 Excellent (can do 40 w.p.m. error free)

5. I think computers are

88

1	2	3	4	5	6	7
Wonderful			(neutral)			Terrible
31	33	14	8	2		

6. Have you used computers in a batch mode for (check all applicable)

91

(1) 7 Have not used them
(2) 39 Information retrieval
(3) ___ Writing programs
(4) 74 Running existing programs
(5) 12 Other (specify) ______

7. Have you specified programs to be written by someone other than yourself?

92

(1) 69 Yes (2) 23 No

Tot. 8. Have you ever utilized a computerized message system, tele-conferencing or computerized conferencing system?

92 (1) 27 Yes (2) 65 No

(If yes, please indicate below which systems you have used).

87 None = 63 Planet-Forum = 2 Arpanet = 2 Confer = 1 Other = 13 2+Others = 5 6 ?= 1

How often have you used computer terminals for: (Check one)

	Never (1)	Occasionally (2)	Frequently (3)
9. Text editing	43	24	24
10. Information retrieval	39	30	22
11. Programming	25	31	35
12. Packaged analysis programs	27	36	28
13. Data entry	27	36	28
14. Game playing	40	39	12
15. Other (specify)	78	2	11

91

Score

7 = 5
8 = 8
9 = 6
10 = 8
11 = 11
12 = 11
13 = 6
14 = 10
15 = 3
16 = 8
17 = 8
18 = 2
19 = 3
20 = 1
21 = 1

16. Have you ever utilized, on a regular basis, a TWIX or like communication system?

Tot.

(1) 13 Yes (2) 74 No 87

17. Please describe your access to a computer terminal at your office or place of work.

(1) 18 No terminal 90
(2) 25 Have my own terminal
(3) 47 Share a terminal

If shared: Own = 23

17a. On the average, how long does it take you to get to the terminal?

2 min = 20 6-9 = 1 ______ Minutes No term = 15 85
2-5 = 24 10-19 = 1 20+ = 1

17b. On the average, how long must you wait for someone else to get off the terminal before you can use it?

______ Minutes

80

Own = 23
2 min = 11 6-9 = 1 20+ = 9
2-5 = 13 10- 19 = 9 No term = 14

18. Do you have a terminal which you keep at home?

Tot

(1)__15__ Yes

91

(2)__51__ No

18a. If no: Is there a terminal available to you that you can take home?

(1)__25__ Yes

(2)______ No

19. What types of terminals do you have access to? (Check all that apply)

89

1) __30__ Hard Copy No hard copy = 9

a) Speed: No speed = 9

__5__ 10 __7__ 15 __50__ 30 characters/second or more

b) Weight: No wgt. = 8

__10__ Under 20 lbs. __27__ between 20 & 40 lbs.

__25__ over 40 lbs.

2) __6__ Visual Display (CRT) No Term = 11

42 both

20. I would not trust computer storage of paperwork that I use daily.

80

__5__ Strongly agree

__21__ Agree

__35__ Disagree

__19__ Strongly disagree

Current Expectations
about the EIES

Tot

1. (a) Concerning the user information brochure about the EIES, check one of the following

89

(1) 3 Did not receive a brochure
(2) 18 Received a brochure, but haven't read it
(3) 50 Found the brochure easy to understand
(4) 8 Found the brochure hard to understand
(5) 10 Read the brochure, but can't evaluate it

(b) Is there any part of the Information Brochure or one-page User's Guide which you had difficulty understanding? (Please be as specific as possible, listing page or section number.) Is there anything that you felt was left out? Any other suggestions about the brochure and/or User's Guide?

2. Which features of the Conferencing System do you anticipate as being most useful to you? (Please rank multiple selections 1,2,3 etc.)

91

Not r'd	Ranked		Ranked 2	3	4 or better or r'd
12	(1) 37	Private messages between individuals	19	5	18
13	(2) 28	Group discussion and conferencing	23	6	21
50	(3) 2	Text editing features	6	12	21
56	(4) 3	Personal notebooks	7	9	16
51	(5) 0	Bulletin	8	9	23
59	(6) 1	Searching the conference records	5	7	19
76	(7) 0	Use of anonymous comments or pen names	3	1	11
87	(8) 1	Other (specify)	1	1	1

3. How much time in the average week do you foresee yourself using the EIES?

88

(1) 8 30 minutes or less
(2) 20 30 minutes to 1 hour
(3) 35 1 - 3 hours
(4) 17 3 - 6 hours
(5) 7 6 - 9 hours
(6) 1 9 hours or more

Tot

4. How often do you foresee yourself signing on the system to send or receive messages or discussion comments?

89

(1)__2__ Once a month or less
(2)__9__ 2 - 3 times a month
(3)__17__ Once a week
(4)__43__ Two or three times a week
(5)__17__ Daily
(6)__1__ Several times a day

5. Do you anticipate entering the material into the System <u>yourself</u> or having someone else do it for you?

92

(1)__64__ Type it myself
(2)__4__ Have it typed
(3)__24__ Both will occur

6. Which statement best describes your incentive for using the System?

91

(1)__3__ I am required to use it
(2)__11__ I have been requested to use it
(3)__17__ I am free to use it as I wish

7. Which of the following best describes your anticipation of the system's worth? (please check only one)

92

(1)__2__ I think it will be useless
(2)__1__ I think it is useful for others, but not for me
(3)__8__ I am skeptical about it but willing to try it
(4)__2__ I am basically indifferent or neutral
(5)__28__ I think it will have limited, but some worth for me
(6)__40__ I think it will be useful in many respects
(7)__6__ I think it will revolutionize my work/communication processes
(8)__5__ It depends (specify) ________________

8. Which of the following do you feel will limit your probable use of the system? (If more than one applies, rank them 1,2,3, etc.)

90

Not r'd		Rank 1		Rank 2	3	4+, r'd
67	(1)	9	Inconvenient terminal location	7	4	3
67	(2)	4	Preference for face-to-face communication	6	5	8
71	(3)	5	Preference for telephone communication	4	2	8
65	(4)	10	The people I wish to communicate with are not on the system	8	3	4
82	(5)	2	Typing skill or lack of a typist	4	0	2
49	(6)	31	Not enough time	3	4	3
71	(7)	4	System too cumbersome or difficult	7	3	5
89	(8)	1	General dislike for computers	0	0	1
77	(9)	1	Prefer drafting by longhand or dictation	3	5	4
72	(10)	7	Other (specify) ________________	4	1	6

9. Compared to the conventional means of communicating with your group, do you expect the EIES to

(1) 25 Involve less of your time
(2) 49 Involve more of your time
(3) 11 Involve the same amount of time

10. How do you think use of EIES will change your communications or work patterns? (Please be specific. What current activities would it replace?)

As 1st Answer

1 ?, little = 15
2 Replace rsrch = 4
3 or nail, phone = 16
4 or spec. activity = 8
5 commen, contact = 29
6 Improve rsrch = 2
7 other = 5
8 no change = 5

11. Why do you personally wish to use EIES? (What do you think you, or your group, or the society, can gain from it?)

12. What disadvantage or negative consequences might possibly flow from your group's use of the system?

13. Any other comments?

14. How long did it take you to fill in this questionnaire? __________

THANK YOU VERY MUCH

Continuation Page

Continuation of Question # ____

Continuation of Question # ____

Continuation of Question # ____

Continuation of Question # ____

Continuation of Question # ____

APPENDIX B
FOLLOW-UP QUESTIONNAIRE, EIES

FIRST FOLLOW-UP

QUESTIONNAIRE FOR USERS OF EIES

INTRODUCTION

TO:________________________________

The questions below relate to your current reactions to the Electronic Information Exchange System, and to possible effects which it may have had upon your work and the development of scientific knowledge within your specialty area. It is the second of three questionnaires which you will be asked to complete for purposes of the overall evaluation of the impact of EIES.

As in all other phases of the evaluation of the EIES system, we will guard the confidentiality of your replies. A copy of your answers will be provided to the evaluator for your group. The data will not be released in an individually identifiable form to anyone else.

There is a continuation page at the end of the questionnaire, for any answers which do not fit in the alloted space. The numbers and spaces in the margins are for use in coding your answers. Because of the "protection of human subjects" regulations, I need to have your "written permission" to take part in this project. Please be sure to sign below and return the questionnaire.

In pretests, completion time averaged only twenty minutes. However, if for some reason you do not wish to complete this questionnaire please check the appropriate space below and return this questionnaire.

Starr Roxanne Hiltz, Ph.D.

Associate Director

Computerized Conferencing &
Communications Center
New Jersey Institute of Technology

________ I do not wish to complete this questionnaire because:

________ I agree to participate in this study

____
SIGNATURE

COL-CODE

I. ACCESS & USE PATTERN

1-4 _____

1. What are the main activities you have been engaging in on the EIES system, and with whom?

Tot.

2. Does anyone else use EIES under your ID? If so, please give their name and approximate on-line time per week.

5 _____

Yes = 16 No = 87 Other time on:

3. In an average week, how many times do you personally "log in" and use EIES? Approximately how long do you usually spend per session?

6-8 _____

9-10 _____

Actual		Preferred
___________	Average # sessions per week	__________
___________	Minutes per average session	__________

11-12 _____

13-14 _____

15-16 _____

4. How much time do you spend "off-line" in an average week doing EIES-related work (preparing entries, filing material received, etc).

___________ 17-19 _____

5. Of the time spent on EIES, what proportions do you spend at your office, at home, or at other locations?.

20-21 _____

22-23 _____

24-25 _____

___________ % at office

___________ % at home

___________ % Other (describe)

100%

COL/CODE

6. COMPUTER TERMINALS
(If you filled out a previous questionnaire and your access to terminals has not changed since then, check here and skip to question 7 on the next page).

Tot.

26 105 — 45 (43%) No change in terminal access since last questionnaire.

Change = 60 (57%)

27 106 — a) Please describe your access to a computer terminal at your office or place of work.

1) 5 (4.7) No terminal — 27 (25.5) No change

2) 40 (37.7) Have my own terminal

3) 34 (32.1) Share a terminal

If shared:

28 103 — On the average, how long does it take you to get to the terminal?

2 = 17 (16.5) Minutes — Has Own = 64 (62%); 2-5 = 12 (11.7%); 10-19 min = 4 (3.9); 20+ = 3 (2.9); No term 3 (2.9)

On the average, how long must you wait for someone else to get off the terminal before you can use it?

29 101 — 2 = 15 (14.9) Minutes — Has own = 70 (69.3); 2-5 = 8 (7.9); 10-19 = 5 (5); No term = 3 (3)

c) Is there a terminal available to you that you can take home?

30 103 — 1) 8 (8) Yes — no change = 29 (28)

2) 27 (26) No — At home = 39 (38)

d) What types of terminals do you have access to? (Check all that apply)

no change = 29 (27.6)

31 105 — 1) 69 (65.7) Hard Copy — no = 7 (6.7)

a) Speed: None = 2 (2) — No hard copy = 35 (35)

32 101 — 4 (4) 10 — 4 (4) 15 — 56 (55) 30 characters/second or more

b) Weight: None = 1 — No term = 35

33 101 — 24 Under 20 lbs. — 17 between 20 & 40 lbs.

24 over 40 lbs.

34 105 — 2) 33 Visual Display (CRT) — No change 30

42 No

COL/CODE

7. Currently, do you yourself type material into EIES, does someone type it in for you, or do both occur? 35 110

1) 92 (84) Type in myself (Answer A below)

2) 5 (4) Have typed in (Answer B below)

3) 13 (12) Both occur (Answer A & B)

A. What type of material do you type yourself? (If more than one, rank-order by frequency). 36 ____

Main	oth.	not r'd	Item	N/A	Code
Main = 76	oth. = 16	not r'd = 2	I type in previously unwritten thoughts/ideas e.g., I compose on line.	N/A = 4	98
Main= 16	oth. = 30	not r'd = 6	I type in rough drafts from outlines or notes.	N/A = 2	54
Main = 8	oth. = 6	not r'd = 14	I type in material that was previously written out and edited.	N/A = 3; 2 or more r'd = 2	33
Main = 2	oth. = 1	2 or more = 1	Other (describe)	N/A = 1	

B. What are the main reasons why you have chosen to have someone else input material for you? (If more than one, please rank-order)

Main	oth.	not r'd	Item		Col/Code
	oth. = 2	not r'd = 15	I don't know how to use the system.	N/A = 91	37 108
Main = 7	oth. = 1	not r'd = 9	I don't have time to use the system myself.		38 ____
2	3	13	I do not know how to type.		39 ____
	2	15	I find using the system directly, i.e., typing at a terminal, incompatible with my professional role or job description.		40 ____
	1	16	I dislike working on line (describe why in the space below)		41 ____
7	2	8	Other (please describe)		42 ____

COL/CODE

8. What do you do with the print-outs of material from EIES?

43 109

1) 4 Throw them all out.

2) 20 Keep them all.

3) 25 Save selective entries in a single file or pile

4) 37 Save selective entries in separate files (please explain filing system below: by subject, author, group, or what).

5) 5 I use a CRT and do not generate print-outs

6) 16 Other (Please describe)

0 = 3	5 = 132	5-10 = 40	11-15 = 11	16+ = 27
(1.4%)	(62%)	(18.8)	(5.2%)	(12.7)

44-45 213 7. How many different people do you feel you are actually exchanging information or communicating with on this system, currently? ______

0=18 (8.5) 5 = 143 (67.1) 5-10 = 45 (21.1) 11-15 = 1 (.5) 16+ = 6 (2.8)

46-47 213 8. Of these, how many have you "met" (gotten to know) over EIES? ______

48 104 9. Have you sent transcripts or other material to persons outside the EIES system, invited other persons to be informal "observers" or otherwise expanded participation beyond your user group? (please explain).

Yes = 54 (52%)

No = 50 (48%)

49 ______ 10. At the present time, which of the following best describes your EIES group?

41 More of a collection of individuals than a research community

54 A set of cliques or subgroups with interests and activities in common, but not an integrated community

3 A well integrated research community that shares many interests and activities in common

COL/CODE

II. OVERALL REACTIONS TO THE EIES MODE OF COMMUNICATION

These questions relate to your overall reactions to the system at this point, as a means of communication and work coordination for your user group. They consist of a number of rating scales on which you are to circle one number which corresponds to where you would place your own impressions of the system on that dimension. For example, here is the first scale:

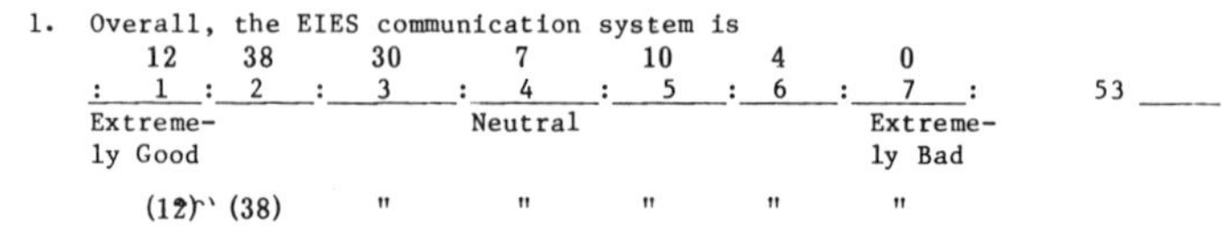

1. Overall, the EIES communication system is

	12	38	30	7	10	4	0	
	1	2	3	4	5	6	7	53 ___
	Extremely Good			Neutral			Extremely Bad	
	(12)	(38)	"	"	"	"	"	

If you think that the system is extremely good, you should circle 1. If you think the system is quite good, you should circle "2"; 3 would mean that the good aspects slightly outweigh the bad aspects. "4", the middle point, should be checked only when the words at the two ends of the scale describe the system equally well. Continuing on, "5" would mean that you feel that the bad aspects slightly outweigh the good aspects, etc.

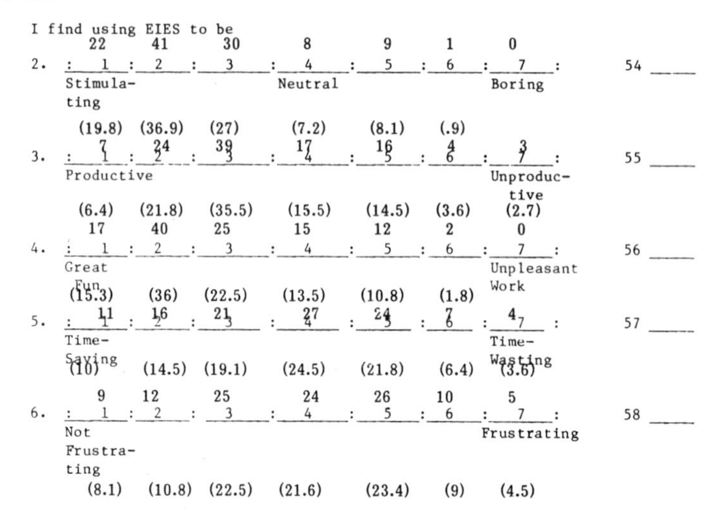

I find using EIES to be

		1	2	3	4	5	6	7	
2.	N	22	41	30	8	9	1	0	54 ___
		Stimulating			Neutral			Boring	
	%	(19.8)	(36.9)	(27)	(7.2)	(8.1)	(.9)		
3.	N	7	24	39	17	16	4	3	55 ___
		Productive						Unproductive	
	%	(6.4)	(21.8)	(35.5)	(15.5)	(14.5)	(3.6)	(2.7)	
4.	N	17	40	25	15	12	2	0	56 ___
		Great Fun						Unpleasant Work	
	%	(15.3)	(36)	(22.5)	(13.5)	(10.8)	(1.8)		
5.	N	11	16	21	27	24	7	4	57 ___
		Time-Saving						Time-Wasting	
	%	(10)	(14.5)	(19.1)	(24.5)	(21.8)	(6.4)	(3.6)	
6.	N	9	12	25	24	26	10	5	58 ___
		Not Frustrating						Frustrating	
	%	(8.1)	(10.8)	(22.5)	(21.6)	(23.4)	(9)	(4.5)	

COL/CODE										MEANS? Std. Dev.	Sig
		18	35	31	16	6	4				
59 _____	7.	: 1	: 2	: 3	: 4	: 5	: 6	: 7 :			
		Friendly						Impersonal		1.2644	.9303
		18	31	24	20	14	3				
60 _____	8.	: 1	: 2	: 3	: 4	: 5	: 6	: 7 :		1.3717	.3227
		Easy						Difficult			

Does using EIES become so demanding of your time and energy that it intrudes upon your capacity to engage in other professional or personal activities?

		16	18	22	24	26	3	1		
61 _____	9.	: 1	: 2	: 3	: 4	: 5	: 6	: 7 :	1.4813	.1110
		Not demand-ing or Intrusive						Very demand-ing or Intrusive		

62 _____ 10. When you send a message over EIES rather than writing or telephoning, would you say that recipients are generally

1) 39 More responsive to an EIES message.

2) 26 Less responsive

3) 33 No difference

3 writing but phoning

63 _____ 1. What is the attitude of your wife, children, or other persons with whom you live towards your use of EIES?

64 _____ 2. Which statement best describes your present reaction to EIES ? (Please check only one)

1) 0 I think it is useless and should be discontinued

2) 4 I think it has its uses for others but not for me

3) 8 I am skeptical but am giving it a try

4) 0 I am basically indifferent or neutral

5) 44 I think that it has certain worthwhile uses for me

6) 33 I think it is very useful in many respects

7) 18 I think it is revolutionizing my work/communications processes.

III. REACTIONS TO SPECIFIC FEATURES OF THE EIES SYSTEM

COL/CODE

1. How valuable or useful do you currently find each of the following features or capabilities of the Electronic Information Exchange System for your own communications activity? (If you have not actually used a feature, please check "cannot say").

	1 Extremely Valuable	2 Fairly Useful	3 Slightly Useful	4 Useless	5 Cannot Say	COL/CODE
Private Messages	66	36	4	1	3	65 ____
Group Messages	29	41	23	6	9	66 ____
Group Conferences	28	42	21	5	13	67 ____
Private Conferences	28	24	8	6	43	68 ____
Public Conferences	16	25	26	9	33	69 ____
Notebooks	14	20	8	7	59	70 ____
The Directory	29	34	20	5	20	71 ____
Retrieval Capability	24	23	14	6	38	72 ____
Text Editing	30	34	14	8	23	73 ____
Anonymity or Pen Name	3	11	17	31	46	74 ____
Explanations	12	31	23	11	31	75 ____
+ SEN	10	12	10	2	69	76 ____
String Variables	13	5	9	2	75	77 ____
News	17	30	35	4	21	78 ____

Comments or suggestions about improving these features or desirable new features? 79-80 ____

COL/CODE

2. Did someone demonstrate EIES to you in person or did you learn from the written materials?

1-4 _____ 1) ___32___ live teacher

5 _____ 2) ___77___ written material only

6-7 _____ 3. How long did it take you to learn to use EIES reasonably well?

1-5 = 184 hours 6-10 = 18 11-15 = 4 16+ = 7

4. Do you now find "How to Use EIES" (on a scale of 1 to 5)

COL							
		51	41	12	3	1	
8 _____	a) understandable	1	2	3	4	5	not understandable
		43	45	7	8	3	
9 _____	b) easy to read	1	2	3	4	5	hard to read
10 _____	c) well organized	1	2	3	4	5	not well organized
		27	43	25	10	2	

11 _____ 5. Suggestions for improvement of the Documentation.

none = 4

Sugg = 43

5a. Do you currently need the users guide (one sheet) or "How to Use EIES" to operate the system?

12 _____ 1) ___31___ User's guide 2) ___16___ "How To" 3) ___37___ Nothing more than 1 = 20

13-14 _____ 5b. If you now operate the system from memory, how long did you rely on the guide to get you through the system? 1-5 = 186 hours 6-10 = 18 11-15 = 3 16+ = 6

15 _____ 6. Have you ever asked a user consultant for help?

1) ___26___ No

2) ___4___ Yes (Please describe whether this was helpful, satisfactory, courteous, or what).

Yes, good = 67 Not reached = 5

Yes, reg. or mixed = 6

7. How would you rate the performance of

COL								
			51	29	18	4	2	
16 _____	Your group leader? (principal investigator)	Excellent	1	2	3	4	5	Poor
			29	25	24	2		
17 _____	Systems monitor (EIES, 100)	Excellent	1	2	3	4	5	Poor

COL/CODE

8. Do you find the language of the system understandable?

	45	43	14	3	3		
a) Understandable	1	2	3	4	5	Confusing	18 _____
b) Courteous	1	2	3	4	5	Inhuman	19 _____
	42	40	16	3	4		

9. (Direct editing commands)

Do you find the use of the +, -, * (special symbols) etc. to be

	34	28	22	11	11		
Easy to remember	1	2	3	4	5	Hard to remember	20 _____
Easy to use	1	2	3	4	5	Hard to use.	21 _____
	39	36	16	7	7		

Comments?

10. Indirect editing commands (.text, .tabs, etc)

Good	1	2	3	4	5	Poor	22 _____
	9	9	20	5	8		

Comments?

11. Which of the following do you currently use to operate the system?

	1 Never	2 Sometimes	3 Frequently	4 Often	
long menu	28	32	15	15	23 _____
short menu	15	38	13	28	24 _____
"answer ahead"	28	26	19	16	25 _____
commands	17	30	16	25	26 _____
string variables	62	20	2	6	27 _____

12. (Answer only if you have used both menus and commands)
Do you now think it is a good idea or a poor idea to introduce the new user to the system through menus, and provide equivalant commands for those who prefer them?

__54__ Good to use menus first

__19__ Should teach commands from the start

5 = Other

13. In EIES, you do not have the choice of permanently refusing to accept a private or group message. Which of the following would you prefer?

__51__ Require acceptance of all messages, as at present
__17__ Require acceptance of private messages only
__26__ Allow rejection of any message, with "message refused by ### returned to the sender

Comments?

1-10
Time to complete (to be corrected)

-9-

COL/CODE

13. Thinking back over your experiences so far with the system, how frequently have you felt..(check one)

COL/CODE	Item	1 Always	2 Almost Always	3 Some-times	4 Almost Never	5 Never
28 ____	Distracted by the mechanics of the System	6 (5.6%)	17 (15.7)	53 (49.1)	25 (23.1)	7 (6.5)
29 ____	Constrained in the types of contributions you could make	4 (3.7)	18 (16.8)	47 (43.9)	30 (28)	8 (7.5)
30 ____	Overloaded with information	5 (4.5)	20 (18.2)	60 (54.5)	18 (16.4)	7(6.4)
31 ____	Able to express your views	26 (24.3)	50 (46.7)	26 (24.3)	5 (4.7)	0
32 ____	Able to get an impression of personal contact with other participants	9 (8.3)	38 (34.9)	50 (45.9)	7 (6.4)	5 (4.6)

14. How satisfactory do you think the system is for the following activities?

COL/CODE	Activity	1 (COMPLETELY SATISFACTORY)	2	3	4	5	6	7 (COMPLETELY UNSATISFACTORY)
33 ____	Giving or receiving information	29 (26.9)	47 (43.5)	20 (18.5)	9 (8.3)	3 (2.8)		
34 ____	Problem solving	6	18	27	27	11	11	1
35 ____	Bargaining	4	16	16	25	11	12	3
36 ____	Generating ideas	24	29	27	10	6	4	3
37 ____	Persuasion	1	13	20	29	20	11	2
38 ____	Resolving disagreements	1	15	24	21	15	15	3
39 ____	Getting to know someone	1	23	31	20	17	8	2
40 ____	Giving or receiving orders	14	20	18	22	10	3	2
41 ____	Exchanging opinions	28 (26.4)	47 (44.3)	25 (23.6)	4 (3.8)	0	2 (1.4)	

PLEASE PLACE A CHECK MARK OR X IN THE APPROPRIATE BOX TO INDICATE WHETHER EACH OF THE FOLLOWING FACTORS HAS BEEN VERY IMPORTANT, SOMEWHAT IMPORTANT, OR NOT IMPORTANT AT ALL IN LIMITING YOUR USE OF THE EIES SYSTEM.

Most Imp.	REASON	VERY IMPORTANT	SOMEWHAT IMPORTANT	NOT IMPORTANT	COL/CODE
					1-4 ____
18 (9.3%)	INCONVENIENT ACCESS TO A TERMINAL	37 (19%)	35 (18%)	123 (63%)	5 ____
3 (1.6%)	RED NOTEBOOK DOCUMENTATION LOOKED LIKE TOO MUCH TO READ	11 (5.6%)	52 (26.7%)	132 (67.7)	6 ____
6 (3.1)	THE SYSTEM IS TOO COMPLICATED	17 (8.8%)	49 (25.4)	127 (65.8)	7 ____
5 (2.6)	TROUBLE WITH PHONE	19 (9.7)	33 (16.8)	144 (73.5)	8 ____
6 (3.1)	TROUBLE TELENET	30 (15.3)	38 (19.4)	128 (65.3)	9 ____
11 (5.7)	COST OF TELEPHONE TELENET	17 (9)	22 (11)	156 (80)	10 ____
14 (7.3)	HAD SOME BAD EXPERIENCES (SYSTEM CRASHED OR DID NOT SEEM TO WORK CORRECTLY)	21 (11)	61 (31)	113 (58)	11 ____
17 (8.8)	LIMITED NIGHT OR EVENING HOURS	38 (19.6)	40 (20.6)	116 (59.8)	12 ____
4 (2.1)	I DO NOT LIKE TO TYPE	10 (5.1)	30 (15.4)	155 (79.5)	13 ____
2 (1)	I DO NOT LIKE USING A COMPUTER SYSTEM LIKE THIS	6 (3)	15 (8)	173 (89)	14 ____
5 (2.6)	THERE IS NO ONE ON THIS SYSTEM WITH WHOM I WISH TO COMMUNICATE A GREAT DEAL	13 (7)	31 (16)	151 (77)	15 ____
3 (1.6)	I AM NOT VERY INTERESTED IN THE SUBJECTS BEING DISCUSSED	11 (6)	33 (17)	151 (77)	16 ____
67 (34.7)	OTHER PROFESSIONAL ACTIVITIES MUST TAKE HIGHER PRIORITY	93 (47.4)	59 (30.1)	44 (22.4)	17 ____
9 (4.7)	THE CONFERENCE COMMENTS OR MESSAGES I HAVE RECEIVED DO NOT SEEM WORTH READING	13 (7)	61 (31)	121 (62)	18 ____
0 (0)	INADEQUATE LEADERSHIP OF THE GROUP	10 (5)	31 (17)	146 (78)	19 ____
23 (11.9)	OTHER (PLEASE DESCRIBE)	39 (57.4)	7 (10.3)	18 (26.5)	20 ____
Tot. 193					

NOW, PLEASE GO BACK AND CIRCLE THE SINGLE MOST IMPORTANT FACTOR. 21 ____

COMMENTS OR EXPLANATIONS?

-11-

V. Conclusion

1. Are there any ideas which you are using or working with at present, which you first learned of on EIES? (Please try to be specific about what you read and what impact it has had on your work).

2. Are you working on any projects or papers at the present time which have been advanced by your use of EIES? (Again, please try to give us some specific details.)

3. Are you coauthoring or collaborating closely with any members of EIES at the present time, using the EIES system? If so, please describe who you are collaborating with, on what, and how you are using EIES in this effort.

4. Are there any "new uses" you have invented for EIES, that are helping you in your work? These uses might not be related to the specific purpose of your group, but we would like to know about them. For example, you might use it to communicate with your family while away on business trips. To coordinate face to face meetings or conferences with other EIES members...

5. Overall, what would you say have been the main negative aspects of use of EIES for your group this far? What things that you wish to accomplish, have not occured, or what undesirable things have occured, that might be attributed to characteristics of communication over the system? Please explain as fully as possible.

6. How long did it take you to complete this questionnaire?________________

   Any additional comments?

Continuation Page

From
question # __________, p.__________

From
question # __________, p.__________

APPENDIX C
POST-USE QUESTIONNAIRE, EIES

POST-USE QUESTIONNAIRE

COL

EIES ID ____________ 1-4 ______

Part I: Your EIES Group's Research Specialty

5-6 ______

Your specialty group is

Name ______________________________

Number ______________________________

1. Is there a commonly accepted "intellectual mainstream" in the specialty?

7 ______

(1) ___51___ Yes (2) ___50___ No

2. To what extent do you feel that you and those with whom you collaborate are in the recognized intellectual "mainstream" of the specialty, or conversely feel you are "isolated" or "peripheral"?
(circle one)

8 ______

Completely in the Mainstream	Somewhat in the Mainstream	Neither in the Mainstream nor Isolated	Somewhat Isolated	Completely Isolated	(no mainstream)
1	2	3	4	5	
16	27	34	15	0	10

How would you rate the degree or intensity of competition within the research specialty?

9 ______

Very Intense	Intense	Moderate	Low	Nonexistent
1	2	3	4	5
2	15	52	29	4

3. What are the reasons for this competition? (Check all that apply).

________ Scarcity of or competition for funds — 10 ______
________ Rival groups of collaborators — 11 ______
________ High achievement or success drive of persons in the field — 12 ______
________ Some persons act unethically — 13 ______
________ Strongly opposing views — 14 ______
________ Other (please describe) : — 15 ______
________ ______________________________

COL

4. Please list the four major or outstanding people in the entire research specialty (not just those on EIES), and the extent to which you currently know them personally and/or are in direct contact with them?

16 ______
17 ______
18 ______
19 ______
20 ______
21 ______
22 ______
23 ______

Extent of Current Contact

			Constant	Frequently	Occasion-ally	Rarely	Never
		a. ______	$\bar{x}$=2.85	13 / 1	22 / 2	24 / 3	21 / 4	7 / 5
(on EIE?)y=		b. ______	$\bar{x}$=3.03	13 / 1	16 / 2	25 / 3	21 / 4	12 / 5
y=49	n=29	c. ______	$\bar{x}$=3.22	7 / 1	15 / 2	24 / 3	27 / 4	10 / 5
y=41	n=32							
y=37	n=3[illegible]	d. ______	$\bar{x}$=3.42	5 / 1	13 / 2	23 / 3	17 / 4	19 / 5

24-25____ 5. Considering all current personal communication modes, what is the total number of different individuals within the research specialty with whom you are currently in contact? $\bar{x}$ ____________

26-27____ 6. How many of these are on EIES? $\bar{x}$=10.78 Media n=6.43
S.D. = 12.7

28 ____ 7. At the present time, which of the following best describes your EIES group?

__42__ (1) More of a collection of individuals than a research community
__43__ (2) A set of cliques or subgroups with interests and activities in common, but not an integrated community
__14__ (3) A well integrated research community that shares many interests and activities in common

29 ____ 8. Has EIES helped to clarify any theoretical controversies in the specialty area?

__9__ (1) yes, a great deal
__39__ (2) yes, somewhat
__52__ (3) no

If yes - please explain briefly the theoretical issue which you think has been clarified through EIES discussions, and the extent to which it has been resolved.

30 ____ 9. Has EIES helped to clarify any methodological controversies in the specialty area?

__4__ (1) yes, a great deal
__36__ (2) yes, somewhat
__59__ (3) no

If yes - please explain the methodological issue which has most benefitted from EIES discussion, and the extent to which you think the issue has been resolved.

COL

Part II Your Work

1. Please list the names of any persons with whom you have co-authored or collaborated in research (colleagues both on and off EIES) since the time you began using EIES. 31___

0 = 23   3 = 11
1 = 20   4-9 = 17
2 = 9   10+ = 4

2. Professional Publications (please try to give exact numbers published in the last year or underway; (means)

	Currently in Progress	Published in Last Year
Text books	.24	.03
Other books	.45	.22
Journal articles	3.1	3.1
Papers presented	2.4 median = 1.3	2.7 (median = 1.5)
Other	.66	1.7

3. How well known is your work, within your specialty area? 52___

	1	2	3	4	5	6	7
	8	15	10	23	24	14	10
$\bar{x}$ = 4.1	Practically unknown			Average			Ranked at top of Field

For the statements below please circle the response which indicates your degree of agreement.

4. Use of EIES has increased my productivity in terms of the quality of work recently completed or underway. 53___

	Strongly Agree	Agree	Neither Agree nor Disagree	Disagree	Strongly Disagree
	1	2	3	4	5
$\bar{x}$ = 3.05	5	33	30	21	14

5. Use of EIES has increased my productivity in terms of the quantity of work recently completed or underway. 54___

	Strongly Agree	Agree	Neither Agree nor Disagree	Disagree	Strongly Disagree
	1	2	3	4	5
$\bar{x}$ = 3.23	5	23	33	27	15

6. Use of EIES has increased my "stock of ideas" that might be used in future work. 55___

	Strongly Agree	Agree	Neither Agree nor Disagree	Disagree	Strongly Disagree
	1	2	3	4	5
$\bar{x}$ = 2.42	19	53	9	13	9

-3-

COL

7. EIES has changed my view of how my own work relates to that of others in my specialty.

56 ______	Strongly Agree	Agree	Neither Agree nor Disagree	Disagree	Strongly Disagree	
	1	2	3	4	5	
	11	37	26	22	6	$\bar{X}$=2.76

8. Participation in EIES contributes to:

a) Short term professional advancement in terms of my current employment

57 ______	Strongly Agree	Agree	Neither Agree nor Disagree	Disagree	Strongly Disagree	
	1	2	3	4	5	
	6	25	30	26	15	$\bar{X}$=3.19

b) Short term professional advancement in terms of my status among my peers in my specialty

58 ______	Strongly Agree	Agree	Neither Agree nor Disagree	Disagree	Strongly Disagree	
	1	2	3	4	5	
	7	35	37	13	9	$\bar{X}$=2.82

c Long term professional advancement with respect to employment

59 ______	Strongly Agree	Agree	Neither Agree nor Disagree	Disagree	Strongly Disagree	
	1	2	3	4	5	
	3	29	37	19	13	$\bar{X}$=3.09

d) Long term professional advancement with respect to my status among my peers in my specialty

60 ______	Strongly Agree	Agree	Neither Agree nor Disagree	Disagree	Strongly Disagree	
	1	2	3	4	5	
	7	35	39	12	8	$\bar{X}$=2.79

9. EIES has provided me leads, references, or other information useful in my work.

61 ______	Strongly Agree	Agree	Neither Agree nor Disagree	Disagree	Strongly Disagree	
	1	2	3	4	5	
	30	51	8	11	2	$\bar{X}$=2.79

10. EIES has increased the familarity of others with my work.

62 ______	Strongly Agree	Agree	Neither Agree nor Disagree	Disagree	Strongly Disagree	
	1	2	3	4	5	
	9	45	34	11	3	$\bar{X}$=2.54

11. EIES has changed my understanding of the interests and/or activities of others in my specialty.

63 ______	Strongly Agree	Agree	Neither Agree nor Disagree	Disagree	Strongly Disagree	
	1	2	3	4	5	
	14	47	27	11	3	$\bar{X}$=2.43

COL

12. How many different people do you feel you are actually exchanging information or communicating with on this system, currently? x=10.7 64-65 ______

13. Of these, how many have you "met" (gotten to know) over EIES? x=5.4 66-67 ______

14. Compared to the conventional means of communicating with your group, has EIES

(1)__36__ Involved less of your time
(2)__48__ Involved more of your time
(3)__11__ Involved the same amount of time 68 ______

15. Has the use of EIES changed the amount of your use of other media in the last year?

Medium	1 Increased	2 No effect	3 Decreased	
telephone	13	63	23	69
mails	20	45	34	70
travel to professional meetings	10	78	11	71
visits with researchers in other locations	12	75	12	72
reading journals or books	28	64	8	73

16. Has the use of EIES affected your communication with any of the following? Colleagues at your institution or organization.

__25__(1) Increased 74 ______
__4__(2) Decreased
__72__(3) No change

17. Colleagues in your specialty but not on EIES

__26__(1) Increased 75 ______
__2__(2) Decreased
__73__(3) No change 76 ______

18. During the year or more that you have been a member of EIES, have you noticed that it has had any impacts on the way in which you think and work, in general?

_____ No 77-78___
_____ Yes 79-80___

If yes-- please describe these impacts in as much detail as possible.

COL

1-4 ______ 19. Communications with researchers in other disciplines or specialty areas

- 45 (1) Increased
- 1 (2) Decreased
- 54 (3) No change

5 ______ 20. Comparing my contributions or effort put <u>into</u> EIES with the amount of information received, I feel that I have

- 5 (1) Contributed significantly more than I have received
- 13 (2) Contributed more than I have received
- 40 (3) About equal
- 23 (4) Received more
- 20 (5) Received significantly more than I have contributed

6 ______ 21. How satisfactory do you think the system is for the following activities? (circle one)

1 = COMPLETELY SATISFACTORY — 7 = COMPLETELY UNSATISFACTORY

COL			1	2	3	4	5	6	7
7	$\bar{x}=2.43$	Giving or receiving information	24	41	13	10	7	3	0
8	$\bar{x}=3.92$	Problem solving	3	15	17	28	22	7	4
9	$\bar{x}=4.13$	Bargaining	5	8	14	25	16	8	8
10	$\bar{x}=2.77$	Generating ideas	15	30	35	8	7	1	3
11	$\bar{x}=4.23$	Persuasion	4	5	29	17	19	15	8
12	$\bar{x}=4.08$	Resolving disagreements	5	8	26	21	16	14	7
13	$\bar{x}=3.25$	Getting to know someone	5	29	33	13	7	7	4
14	$\bar{x}=3.21$	Giving or receiving orders	9	33	13	17	7	5	6
15	$\bar{x}=2.34$	Exchanging opinions	25	41	19	5	5	1	2
16	$\bar{x}=3.30$	Expressing <u>positive</u> information	7	24	32	15	8	4	6
17	$\bar{x}=3.54$	Empressing <u>negative</u> emotions	7	22	20	21	16	5	5
18	$\bar{x}=3.86$	Sociable relaxation	2	19	25	21	12	7	10

22. Please estimate the maximum you would pay for EIES under the conditions described and how much you would use it.

COL

(medians, including ZEROS)

	Cost in Dollars per Hour	Hours of Use Per Week	COL
EIES with current membership if a) Financed from your pocket	$2.40		19-20 ______ 21-22 ______
b) Financed by another source	$6.38		23-24 ______ 25-26 ______
EIES with peer group of your choice, if a) Financed from your pocket	$3.58		27-28 ______ 29-30 ______
b) Financed by another source	$8.50		31-32 ______ 33-34 ______

23. What one or two factors best explain why you have not used EIES more?

35 ______

36 ______

37-38 ______

24. How many hours do you feel it took you

a) To learn the basic mechanics of sending and receiving messages and comments ______ hours (median = 1.84) 39-40 ______

b) To feel comfortable communicating with others using this medium ______ hours 41-42 ______

c) To learn the advanced features which you wanted to use ______ hours 43-44 ______

III. Reactions to Specific Features of the EIES System

1. How valuable and useful do <u>you</u> currently find each of the following features or capabilities?

	Frequency of Use			Value				
	(1) Frequently	(2) Occasion-ally	(3) Never Used	(1) Extremely Valuable	(2) Fairly Useful	(3) Slightly Useful	(4) Useless	(5) Cannot Say
Private Messages	69	28	2	67	20	10	0	1
Group Messages	22	57	19	35	27	26	2	6
Private Conferences	23	41	33	33	24	8	4	28
Group Conferences	44	37	16	36	31	14	2	13
Private Notebooks	6	32	58	13	23	6	5	48
Group Notebooks	3	18	73	7	15	7	5	62
The Directory	16	68	13	32	33	16	3	10
Chimo	27	42	27	17	23	22	5	29
Retrieval of items already read	11	60	27	30	29	9	3	26
Searches for items	5	52	39	27	15	17	1	35
Voting	11	83	11	2	12	8	1	73
Use of ?;??	6	40	44	11	22	15	4	38
Explanation File	3	38	44	9	17	17	4	38

45 ____
46 ____
47 ____
48 ____

49 ____
50 ____

51 ____
52 ____

53 ____
54 ____

55 ____
56 ____

57 ____
58 ____

59 ____
60 ____
61 ____
62 ____

63 ____
64 ____
65 ____
66 ____
67 ____
68 ____
69 ____
70 ____
79-80 = 09

	(1) Frequently	(2) Occasion-ally	(3) Never Used	(1) Extremely Valuable	(2) Fairly Useful	(3) Slightly Useful	(4) Useless	(5) Cannot Say	1-4=ID
Synchronous discussions in conferences	2	22	72	8	11	16	2	60	5 ___ 6 ___
System commands (e.g., +cnm)	34	36	26	37	24	7	1	24	7 ___ 8 ___
User defined commands (+Define)	3	23	70	20	15	4	0	59	9 ___ 10 ___
Anonymity or Pen Name	6	22	67	10	12	16	12	45	11 ___ 12 ___
User consultants and/or HELP(110)	12	63	21	47	19	7	3	19	13 ___ 14 ___
Text editing (direct) (e.g;/old/new/;*)	45	29	21	49	19	2	1	23	15 ___ 16 ___
Text editing (indirect) (e.g.; text)	12	20	63	20	17	2	0	57	17 ___ 18 ___
Games (e.g. +story)	1	32	63	3	6	19	11	57	19 ___ 20 ___
Special programs (e.g. +terms; +respond)	1	19	76	9	9	7	0	72	21 ___ 22 ___
Interact programming	1	5	88	5	3	7	0	80	23 ___ 24 ___
Terminal Control features (e.g. +Left, +page;+slp)	5	25	66	10	18	6	0	63	25 ___ 26 ___
Graphics routines	3	92	10	7	5	2	1	81	27 ___ 28 ___
+SEN and ???	15	25	55	18	19	8	2	50	29 ___ 30 ___
Tailored Interfaces (e.g., + Legitech)	8	75	22	3	5	3	0	74	31 ___ 32 ___

2. Are there any particular features of EIES you have found to be (Please describe and comment)

a) Unique and valuable to this type of system?

b) Useless, distracting and/or out of place in this type of system?

c) What general improvements/new features/changes would you like to suggest for EIES?

3. EIES is now at the stage where certain individual users and groups are constructing specialized interfaces and data structures. Do you now see any particular items of this nature that would have been particularly beneficial to your group?

4. Any other comments on the EIES system or its impacts, or on this questionnaire?

THANK YOU!!!

APPENDIX D
RESPONSE RATE, EIES QUESTIONNAIRES

Group	30	35	40	45	54	Tot	Rate
Pre-Use							
Sent	30	35	40	35	35	175	
Returned	15	23	32	22	8	98	56%
Short F-up							
Sent	10	14	20	31	26	101	
Returned	9	12	16	26	12	75	75%
Long Follow-up							
Sent	26	35	37	27	15	140	
Returned	22	24	30	21	9	106	69%
Post-Use							
Sent	25	30	42	30	30	157	
Returned	19	24	31	19	8	102	65%

APPENDIX E
SAMPLE USER CONSULTANT FILE REPORT, EIES
SUMMARY OF USER INQUIRIES FOR FEBRUARY, 1978

During the month of February, 141 interactions between user consultants and users were logged. This log includes the problems addressed to the user consultants and the responses to them. The log was established by Roxanne Hiltz to serve as an unobtrusive way of collecting data on user problems, out of which data could emerge a basis for making decisions regarding the nature of and priority of improvements in documentation and other features needed for the EIES system.

The main problems encountered are similar to those of earlier months:

1. There were fifteen problems with the use of the various commands for copying in and out of the scratchpad (&<M12345, +cy C29C40, +cya n104 p28, etc.). This material is considered an "advanced feature" and is not described in "How to Use EIES." However, since various versions of the system were initiated during the month, even experienced users were caught unawares by the changes in specifications, such as whether or not the @ sign should be included in a command.
2. A related problem involved seven requests on how to use the storage areas. Their usage is briefly described in the user materials; unfortunately, the examples given do not work with current versions of the system.
3. Eight more new users reported the "mysterious problem of double printing." More instructions telling how to set the terminal for half duplex and informing the user that double printing means that something is not set for half duplex need to be included in the next revised version of the basic user materials.
4. There were ten problems with the use of notebooks, which are a feature not specifically documented in the existing user materials. Involved in these were five new users who assumed that one gets a personal notebook automatically. One suggestion is that either Murray Turoff or the System Monitor send a message, waiting for all members when they first sign on, instructing them how to request a personal notebook or conference from the System Monitor.
5. There were several users at the beginning of the month complaining that they did not know how to find out what conferences were going on in the system. One of them sent a marvelous description of the "Catch 22" situation:

KEYS:/I WANT TO JOIN/

IN ORDER TO GET MESSAGES FROM A CONFERENCE, YOU MUST BE A MEMBER.

IN ORDER TO BECOME A MEMBER, YOU MUST GET THE OK OF THE CONFERENCE MONITOR.

IN ORDER TO FIND OUT WHO THE MONITOR IS, YOU MUST QUERY THE SYSTEM ABOUT THE CONFERENCE.

BUT IN ORDER TO QUERY THE SYSTEM ABOUT A CONFERENCE, YOU MUST BE A MEMBER.

THUS IN ORDER TO JOIN A CONFERENCE, YOU MUST ALREADY BE A MEMBER.

***************UGH!!!!***************

I WOULD LIKE A LIST OF ALL THE CONFERENCE TITLES AND A LIST OF CORRESPONDING CONFERENCE MONITORS SO I CAN ASK TO JOIN THOSE THAT LOOK INTERESTING. spacel

This problem was resolved by setting up Public Conference 1008 for a listing and description of all conferences on the system that others may ask to join and by having the group moderators send out messages to their groups reminding them of the various conferences and moderators.

6. There were eight "bug" reports, which were referred on to the programmers.

No other problems were reported more than twice.
Resolution of problems one and two is now being discussed.

APPENDIX F
EXAMPLES OF COMPUTERIZED REMINDERS AND THANK YOUS

1. REMINDER MESSAGES

****** A GENTLE REMINDER******

I have not yet received your follow-up questionnaire.
If it is in the mail, thank you.
If you have not received one or need a new one, please message me.
And if it is just lying around, won't you please take about twenty minutes and fill it out?

Anxiously yours,
Roxanne

CHRISTMAS REMINDER

HI!
While
we know that
you did not intend
to send EIES anything
more than Holiday Greetings
at this time of year- if even that-
you can easily give us a very wonderful
present costing only a little of your time:
FILL OUT YOUR YELLOW POST-USE QUESTIONNAIRE
AND MAIL
IT BEFORE
CHRISTMAS
*P*L*E*A*S*E*

HAPPY HOLIDAYS!

ON-LINE THANK YOUS

Presented in Appreciation of your Outstanding Questionnaire-Completion Efforts

```
TTTTTTT  AAAAAAA        DDDDD    AAAAAAA  !
   T     A     A        D    D   A     A  !
   T     AAAAAAA  ===   D     D  AAAAAAA  !
   T     A     A        D     D  A     A  !
   T     A     A        D    D   A     A
   T     A     A        DDDDD    A     A  !
```

```
CCCCC  OOOOO  N    N  GGGGG  RRRRR  AAAAA  TTTTT  SSSSS  !
C      O   O  NN   N  G      R   R  A   A    T    S      !
C      O   O  N N  N  G  GG  RRRRR  AAAAA    T    SSSSS  !
C      O   O  N  NN  G   G  R R    A   A    T        S
CCCCC  OOOOO  N    N  GGGGG  R  R   A   A    T    SSSSS  !
```

YOU HAVE COMPLETED YOUR LAST QUESTIONNAIRE! WELL DONE!
YOU can feel very proud; you are an EIES member in good standing.
WE can relax; we have your data. Thank you.

*   *

APPENDIX G ONE PAGE USERS GUIDE TO EIES

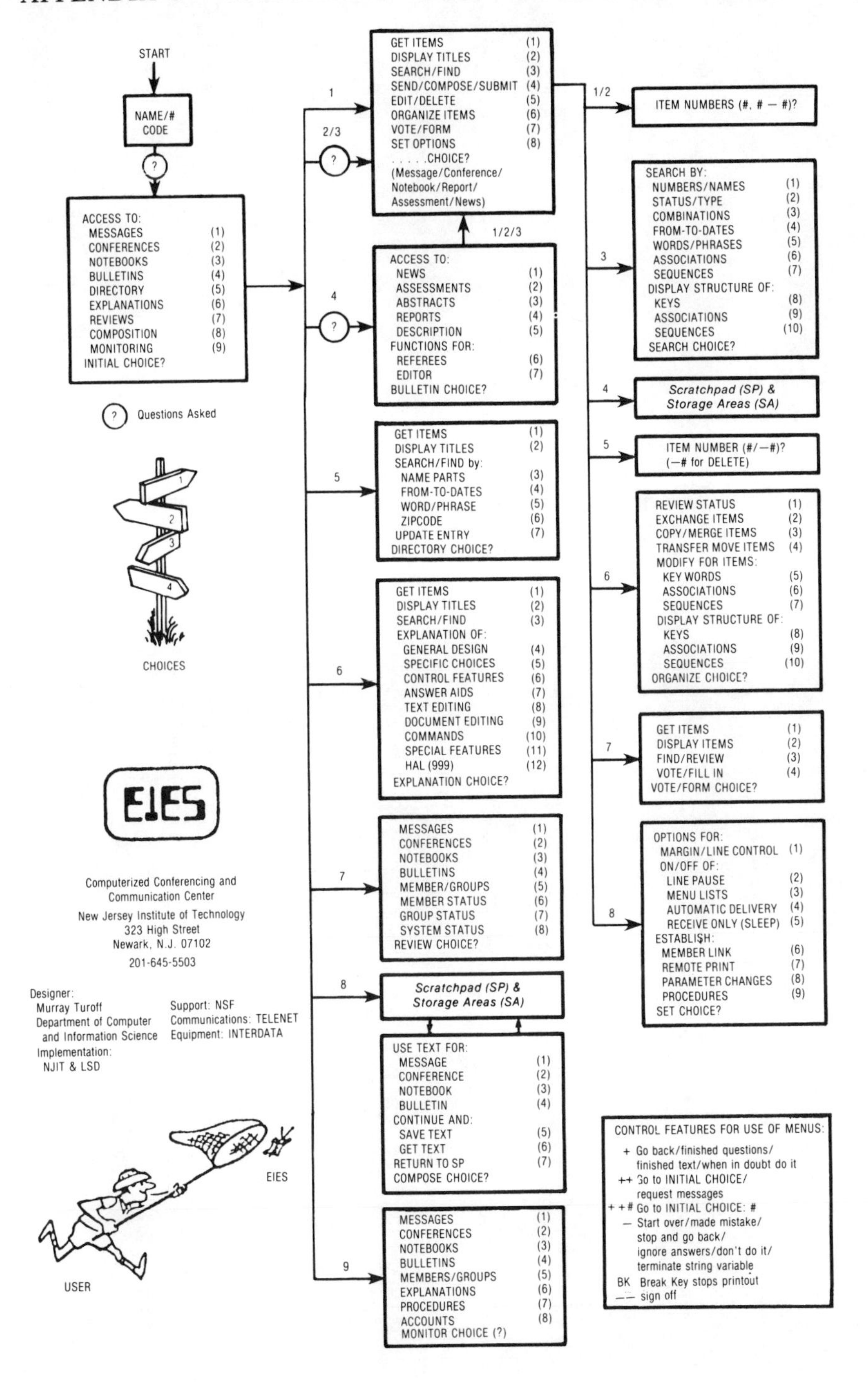

AUTHOR INDEX

J

K

L

M

N

O

P

R

S

T

U

V

W

Z

SUBJECT INDEX

T

U

W